Ornamental Plants and Flowers of Tropical Mexico
- a guide for the curious tourist or resident

"Residents and visitors alike will appreciate and benefit from Linda Abbott Trapp's Ornamental Plants and Flowers… Her photos and well-researched information make it fun and easy to identify the commonly encountered plants of the region. The book helps fill the void for those curious to learn more about the many fascinating plants, trees, and flowers of Mexico."
-Bob Price, Curator, PV Botanical Gardens

"A few years back, when I first set foot in Vallarta, I was amazed at the diversity of tropical plants and also amazed that no one had written a book like this. This new book, which I have been pleased to help edit, is an essential tool for the tourists and locals. With it, they will become better acquainted with the marvelous plants and trees that are the essence of Mexico's tropical allure."
-Daniel Lamarre, Writer, Journalist and Tropical Plants Specialist

Text and Photographs by
Linda Abbott Trapp

Printed in Korea   Published in Puerto Vallarta, Mexico
Copyright © 2006, Linda Abbott Trapp

Cover design by Ann Marie Danimus
Venegas Realty photographs by Fernando Castillo, NH Krystal pool and façade, Ruben Hidalgo
Cover photos: right, Ginger, left, top to bottom, Water lily, Sago palm, Heliconia, Canna lily.
Back, Royal palms. Title page, Hibiscus, this page, Sago palm. Acknowledgments, Livistona chinesis, How to Use this book, (left to right) Thunbergia, Fishook heliconia, Cordyline, Portulaca. "The Plants and Flowers", Royal Palms. Sponsoring Hotel and Development Gardens page top, Manilla palms. Bibliography, Water lilies.
All rights to text and photographs reserved under International and Pan-American conventions.
No part of this publication may be reproduced, stored in a retrieval system, or transmitted in any form or by any means, electronic, mechanical, photocopying, recording, or otherwise, without prior written permission of the author.
ISBN number: **1-59971-252-0**

Dedicated to those who
Work to preserve the beauty of the earth
And those who labor to teach children the value
And wonder of their environment

Introduction

Preparing this book has been a labor of love, a true amateur delight. It first arose out of frustration, for I couldn't find a similar book to answer my questions, and those of our many visitors here in Puerto Vallarta. It grew to become an all-consuming joy for these last months. Because of my academic background, the research has been fascinating, and the disagreements among experts on nomenclature and care, familiar. Because of my background in the arts, the process of photography and design has been absolutely delightful and fulfilling. The primary challenge has been what to omit of the incredibly beautiful, graceful, and occasionally, odd flowers, plants and trees that are everywhere in this tropical wonderland.

The book is arranged alphabetically, by what I found to be the primary name of the plant. The index includes for many of the plants here, more than one name. In addition to the English and Latin names, natives and visitors have a variety of names for many of these plants and flowers, and some of the vernacular names they declined to tell me for reasons of politeness. I've chosen the most familiar as the primary name, and am more than ready to make adjustments as needed should another edition be in the works after all my errors come to light. Similarly, of the roughly 50,000 species of plants in Mexico, only a few more than a hundred appear here. For the most part, I've chosen plants closely associated with tropical Mexico in the visitor's mind, those commonly used in resort landscaping, a few imported plants I was especially charmed by, and some of those plants the visitor or resident might see on a tour off the beaten path as well. There is clearly room for many an additional reference to supplement this small start. My hope is that this slim volume will add to your joy, cause you to more fully celebrate life on this wondrous planet, and answer at least a few of your questions.

How to Use this Book

If you've the leisure and interest, you might choose a relaxing spot and simply flip through the book, looking for photographs of plants you've wondered about. If you prefer a more efficient process, the index lists the common name, Latin name, and alternate names I've discovered for each entry. The book is alphabetical to make the most of the index cross-references. My intent has been to provide some information on use and cultivation, with the occasional note of interest as I became aware of such items. For further detail, there is a brief bibliography. In addition, the reader is encouraged to explore the many online resources, exercising some caution as to accuracy.

Acknowledgments

In this case, it is even truer than for most books, that I could not have done this myself. For starters, botany is not my field, only a new-found passion. Secondly, although researching plant information is feasible with internet resources and a handful of really good books, it's essential to first have the name of the plant. Quite often, when I found an interesting or especially lovely specimen to photograph, I didn't know its name. Nearly always, someone nearby would try to help by providing a name, and just as frequently, that name and others I was told for the same plant were quite different. The appreciation and enjoyment of the plants and flowers was, however, universal, consistent, and heartfelt.

I owe a debt of gratitude to the sponsors, the NH Krystal, the Four Seasons Resort, and Venegas Realty, Real Estate and Investment Services. Members of the Puerto Vallarta Writers Group, especially Daniel Grippo and Twila Crawford, have been most encouraging throughout the writing process, and writer Bob Rossier was unflagging in his practical support. My husband, Robert Trapp, deserves kudos for his patience and willingness to go on innumerable photo shoots. For botanical expertise, I was fortunate to be introduced to some wonderfully knowledgeable enthusiasts. Bob Price of the Botanical Gardens of Puerto Vallarta and Alejandra Quintero Coello of Arte Verde, near Mezcales, answered many, many questions, as did the staff at the Jardin Botanico in Mazatlan. Felix and Adelien Montes were kind enough to permit me to photograph their marvelous palm ranch, Tropical America, in Bahia de Banderas, and some of the loveliest settings represented here are a result of their work and dedication. Paul Sanders and Kathleen Dobek identified several plants for me. Bill and Bobbie Snyder shared their beautiful garden, Linda Joy and Barbara Poindexter made thoughtful design comments. John Isenhower provided solutions to an unbelievable number of computer-related questions. Daniel Lamarre brought his years of experience with tropical plants, his passion for the field, and his editorial precision to bear on the work, and help me overcome the most egregious errors. The remaining mistakes are simply mine.

Sponsors:
- NH Krystal
- Four Seasons Resort
- Venegas Realty
  - Real Estate and Investments

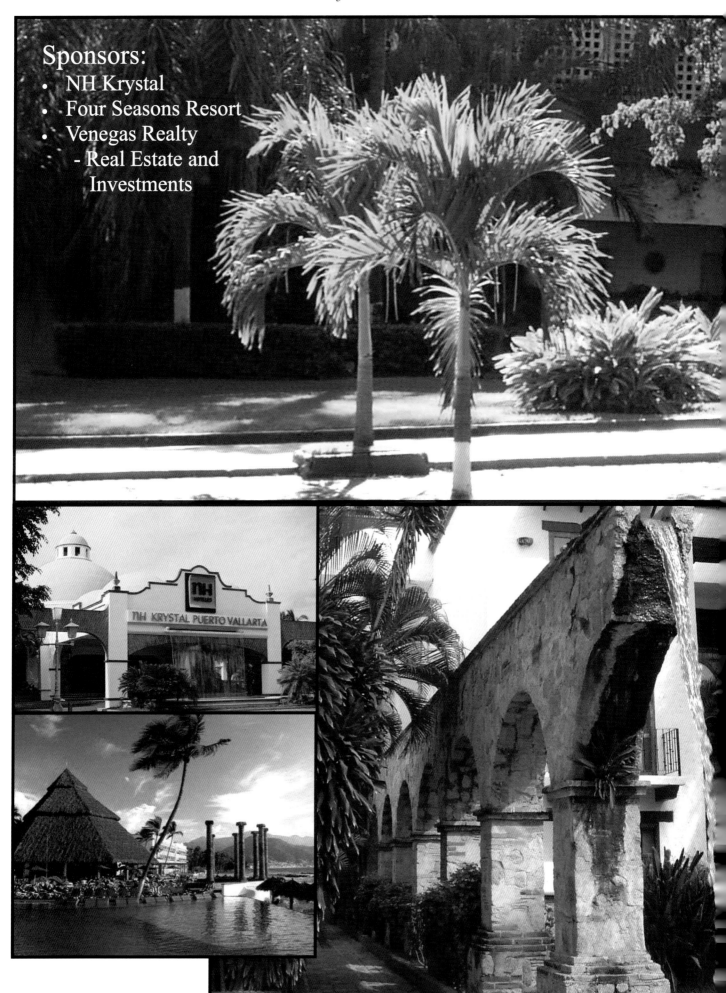

FOUR SEASONS RESORT
Punta Mita, México

The Plants and Flowers...

Acalypha (Amarantha) (*Acalypha hispida*)

Family: Euphorbiaceae
Alternate Names: Beefsteak Plant, Red Hot Cat's Tail, Fire Dragon Plant, Copper Leaf

Use: Eye-catching hedge or background plant, including the *A. hispida* (photograph) and the *A. wilkesiana*, a tall shrub with variegated leaves related to the poinsettia. The copper leaf variety has smaller tassels and variegated leaves in combinations of pink, red, green, and white. Can be used as houseplants.
Flowers: *A. hispida's* wooly red flower spikes are chenille-like tassels, and may droop to 18 inches on mature plants.
Cultivation: Full sun to part shade. Moist soil, fertilizer twice monthly. Provide bright light for Acalpha if grown as a houseplant.
Propagation: Cuttings root easily.

Allamanda (*Allamanda cathartica*)

Family: Apocynaceae
Alternate Names: Golden Trumpet, Bush Allamanda

Use: Grown as an informal fence in tropical climates, as houseplants elsewhere. Excellent in planter boxes, especially when used with other brightly colored flowers for contrast.

Flowers: Clusters of showy, trumpet-shaped flowers, to 4" in diameter. *A. blanchetii* is purple, the more common *A. cathartica* is jonquil-yellow. The variety, Stansill's Double, has double yellow flowers. In cultivation, generally does not produce fruit.

Cultivation: Vining or loosely shaped shrub in habit. Requires full sun, warm temperatures for best flower production. Water lightly in the cooler months, and provide more water during hotter periods. Fertilizer, especially liquid manure, produces vibrant flower clusters. Prune heavily to improve shape.

Propagation: From 3" tip cuttings taken in the Spring. The reddish-purple type is frequently grafted onto rooted cuttings of *A.. cathartica.*

Note: Mildly poisonous, used in very small quantities as a purge.

Anthurium (*Anthurium*)

Family: Araceae
Alternate Names: Little Boy Flower, Flamingo Flower, Obake, Palette Flower, Spathe Flower

Use: Staple of the florist trade, these vivid, waxy flowers are highly decorative in exotic arrangements, often combined with orchids, although not related. Have been used as ground cover in tropical landscaping, creating a stunning, although not particularly practical, effect. There are over 700 species.

Flowers: The spadix, or column, consists of a multiplicity of tiny flowers, which is protected by the colorful decorative shield or spathe. Cut flowers are extremely long lasting, and blooms on a growing plant can last for months.

Cultivation: Prefers partial shade, requires consistently high humidity, a rich, moisture-retaining soil, and excellent drainage. They can be cultivated indoors near a window in a sunny heated room, with best results obtained in humid greenhouses.

Propagation: by seeds, cuttings, or division.

Bamboo (*Arundinaria*)
Family: Gramineae

Use: Wide-ranging, primarily including building materials, food and food preparation utensils, musical instruments, fishing poles, and handicraft. Its use was recorded as long ago as 1000 B.C. in India. Commonly planted as an ornamental. There are about 1,000 species, including green, black, and yellow forms. Some, especially golden bamboo, can be invasive in landscape uses. Photos below, right, show African bamboo, *B. vulgaris* is at left.

Flowers: Once every several decades, all the bamboo plants in an area will flower at the same time, producing many flowered spikes. In between these displays, there are no flowers at all.

Cultivation: Grows to 60' high and 4" in diameter, and is often grown and managed in plantations as a renewable resource. Needs fertile, moist soil in cool places. Forms large clumps. The fastest-growing types are found in Asia. Many types of bamboo are resistant to disease and to insects, and can be easily grown without intensive labor.

Propagation: By offshoots for some types, and cuttings of various types may be taken from mature plants.

Banana (*Musa*)
Family: Musaceae

Use: This tree-like plant (actually a herb), closely related to plantains, produces edible fruit which grows in clusters. There are many fruits to each tier, which is called a hand. Bananas are the staple starch in some tropical populations. Bananas are eaten raw, cooked, fried, or dried.

Flowers: Orange-yellow in color, formed on long drooping stalks. The tender inner flowers are used in cooking, primarily in Asia.

Cultivation: Fast-growing perennials that can reach tree height in a single growing season. Cold-hardy varieties make handsome ornamentals in cooler climates. Full sun needed for best growth, and shelter from excessive wind. Needs a great deal of water and fertilizer, and deep mulch. Prune ornamentals for looks after any cold season. Can be grown as houseplants or in containers outside.

Propagation: Seedless varieties are propagated asexually from offshoots. Plants seem to move, or "walk", over time as the lateral rhizome formation dictates. Prune unwanted offshoots.

Bird of Paradise (*Strelitzia reginae*)
Family: Musaceae
Alternate Names: Crane Flower, Crane's Bill

Use: The large and showy flowers are commercially grown as a staple of the cut flower business. The plant works well in formal arrangements, as a bedding component, and as a prominent feature of open tropical garden spaces. Often planted near pools and water features, since it produces very little litter.

Flowers: Large and exotic, the orange, blue and purple flowers resemble tropical birds, hence the name. Flowers continuously throughout the year, with one orange flower emerging each day from the green or purple boat-shaped 9-11" bracts.

Cultivation: Does well in moist, tropical gardens, but also is quite drought resistant. Needs full sun. Can be grown under glass in cool climates. This plant is grown in frost-free climates world-wide. Will grow well in good garden soil. In containers, needs an enriched mixture of loam, peat, and, sand. Part shade produces more attractive leaves, but fewer flowers. Will grow to 5' in height.

Propagation: The bird-like appearance attracts birds, which helps with the spread of pollen. Propagate by seed, or by division of rooted suckers.

Note: The scientific name was given in honor of the wife of King George III, Queen Charlotte of Mecklenburg-Strelitz.

Bismark Palm (*Bismarckia nobilis*)
Family: Areceae

Use: Landscape specimens, singly or in rows. Height from 40-70', reaching 15' across, with a crown of spiky fronds.
Individual leaves may be more than 36" across.
Cultivation: Sun or part shade, drought tolerant, salt resistant. Does not need fertilizer once established. Likes sandy soil. Does not attract pests. Prune with care, avoiding damage to lower trunk. Does not transplant well, due to deep taproot.
Propagation: From seed.

Blue Agave (*Agave Tequilana Weber Azul*)
Family: Liliaceae
This succulent (not a cactus) is related to the lily and the amaryllis.

Use: There are 250-300 different agaves, grown from the southwest U.S. to the South American tropics. Of these, the Blue Agave, shown, is used solely in the production of tequila, made from the heart, or pina. The remainder of the plant has no other use, although they are at times used in decorative plantings due to their striking appearance. The records indicate the Aztecs had a multitude of uses for the Blue Agave, including soap, clothing, and string.

Flowers: In the wild, the pina (core) sends up a shoot that bears small yellow flowers at its top.

Cultivation: Blue agave grows best above 5000 feet, where the mature plant at about 8 years may have leaves 5-8' tall and be 7-12' in diameter. The plant is watered only by rainfall during the rainy season, and lives about 8-15 years. Some say it grows best on volcanic slopes. Pruning the points of the leaves encourages the pina to grow larger. The pina may reach a weight of 500 lbs, but most average 200 lbs at harvest. Hand weeding is often used, but some growers may spray with pesticides.

Propagation: Grown from shoots taken in the beginning of the rainy season during the 4th-6th year.

Bottle Palm (*Hyophorbe lagenicaulis*)
Family: Arecaceae

Use: Because of its unusual shape and moderate size, 12'-15', this is a novel choice for landscape interest. The grey trunk grows to 2' in diameter and is swollen into a bottle-like shape. The canopy of 4-8 elegant leaves, each 9-12' long, looks striking.
Flowers: Cream male and female flowers appear from the same inflorescence. The flower stalks come from below the crownshaft, and are followed by black oval fruit of 1.5" length.
Cultivation: Semi-desert in origin, this slow-growing palm thrives in sun to part shade. It is highly salt tolerant, and can withstand some drought, although for best growth it requires regular moderate water. Avoid overwatering.
Propagation: From seed. Transplant after one year.

Bougainvillea (*B. Scarlett O'Hara*) (red)
Family: Nyctaginaceae
Alternate Name: Paper Flower

Use: Ornamental climber for covering fences and walls; also seen as a sprawling shrub. Can be grown in pots, or trimmed as border hedge. Bright green heart-shaped leaves, also variegated varieties, thorns.
Flowers: Exuberant clusters, comprised of small white flowers surrounded by larger, papery, colorful bracts. Amount of sun affects color; lighter varieties profit from some shade. Double-flowered types hold blooms even after fading. Common colors include cerise, orange, white, pink, and mixtures of these.
Cultivation: Rich soil, water moderately during bloom period, full sun except for white and yellow varieties. Prune after flowering.
Propagation: By stem cuttings. Tie shoots to sturdy support.

Breadfruit (*Artocarpus altilis*)
Family: Moraceae

Use: The breadfruit has been a mainstay of some tropical cultures, such as the Polynesian, where the wood is used for canoes, the bark beaten into cloth, the sap from the gum used for everything from caulking to catching birds. Of course, the fruit is eaten baked, boiled, pickled, or pounded into a paste. The large, beautiful, deeply lobed leaves make the tree a desirable landscape specimen. They turn tan before falling, and hold their shape well for arrangements.

Flowers: Numerous and small, the male flowers are in spikes, the female, in dense masses on the same tree. The fruit is rounded, green, and large. Some cultivars bear seedless fruit, but the most common has large seeds.

Cultivation: This large (up to 80' or more) tree grows in most wet tropical countries. It is found in full sun. It needs a hot climate, and fruits most prolifically in the rainy season.

Propagation: From cuttings, or from the shoots which grow upward from the roots. Can also be started from seed.

Note: This is the fruit of *Mutiny on the Bounty* fame, which Captain Bligh thought would be appropriate food for the slaves. Caribbean cookbooks carry recipes that call for breadfruit today.

Bromeliads
Family: Bromeliaceae

Use: Only the Pineapple is used for food, but several other of the 3000 bromeliad species are sources of fiber. They are widely used as ornamental plants, growing from sea level to about 8000' altitude. They are prized for their ease of cultivation, brilliant long-lasting flowers, and beautiful foliage. Some, like *Aechmea trifasciata* (upper right), produce berries.

Flowers: Flowers appear at maturity and are seasonal by species. Ethylene gas seems to trigger the flowering. Many have inconspicuous true flowers surrounded by showy bracts.

Cultivation: Some are true epiphytes, or air plants, others, terrestrial. They grow on rocks, beaches, on other plants, or in the forest understory. They are found in temperate to tropical climates, and are widely grown indoors. All have a spiral rosette of leaves. Depending on species, they may require full sun to full shade, but all prefer free draining soils to avoid danger of rot.

Propagation: From pups or seed.

Butterfly Palm (*Dypsis lutescens*)
Family: Arecaceae
Alternate Names: Golden Cane Palm, Yellow Palm, Golden Feather Palm, Madagascar Palm.

Use: Attractive landscape plant, slender, with many trunks, and with age, multiple crowns. It is also popular as an indoor plant but somewhat difficult to grow there. Shrub-size in cool climates, grows to 30' in warmer climates.
Flowers: Flowerstalk emerges from below the crownshaft. Yellow flowers are followed by fruit which ranges from pale yellow to nearly black.
Cultivation: Fast growth, requires thinning excess suckers to create a more attractive, open, cluster. Needs well drained rich soil, kept moist. Flourishes in sun or part sun. Shallow-rooted, fairly salt-tolerant. Stalks are yellow-green, Leaves are pointed in contrast to the similar MacArthur Palm.
Propagation: From seed, or use the suckers detached in thinning to create new clumps.

Buttonwood (*Conocarpus erectus*)
Family: Combretaceae
Alternate Name: Button Mangrove

Use: Both the green and silver (*C. erectus sericeus)* varieties are available as ornamental trees, and have been cultivated as bonsai. The heavy wood is durable, taking a fine polish, and is used for boats and maritime construction, as well as for pilings and firewood. The bark is used for leather tanning, and the leaves and other plant parts have medicinal uses as well.
Flowers: The many small cuplike flowers are greenish, have 5 lobes, and appear in a ball shape. The fruit clusters are red to purple, and look very much like frosted raspberries.
Cultivation: This wide-ranging tree with its spreading crown grows in a variety of tropical and subtropical climates, usually in brackish or saline silts, on shores, marshes, and stream banks. It tolerates salt and a range of temperatures and abundant or scarce water. It grows in wet soil and in hard packed sand.
Propagation: From seed, with high germination rates.

Calabash (*Crescentia cujete*)
Family: Bignoniaceae
Alternate Name: Tree Gourd

Use: Its dried fruits are used for ornamental purposes, and the hard shell is made into maracas, domestic utensils, and pipes for smoking. The wood is tough and flexible.

Flowers: The night-blooming flowers are pendulous, solitary, yellow to purple, and grow on long, rope-like hanging stalks. They are pollinated by bats. The flowers are followed by many-seeded, hard-shelled gourds that grow as large as 14" diameter. Both flowers and gourds grow from nodes on the trunk and branches.

Cultivaton: This small-to-medium tree (to 30') grows in open grasslands. Its wide-spreading horizontally divided branches have leaf clusters at intervals. Fruits take up to 6 months to ripen. Some experts report that the Calabash will survive a mild frost; others disagree.

Propagation: From seed or from cuttings..

Caladium (*C. x hortelanum*)
Family: Araceae
Alternate Name: Geraniums of the South

Use: Bedding or container plant grown for its colorful foliage. There is some debate about its 7-17 varieties, their names and origins, but all are beautiful with arrow-shaped leaves with colorful patterns or blotches.
Flowers: In a spadix surrounded by a yellow-green spathe.
Cultivation: Found in the wild from sea level to over 3000' elevation, this understory herb has adapted well to landscape and indoor cultivation. Native to tropical climates, these plants are cold sensitive. They are shade plants, although part sun will do, and they prefer a well-drained soil kept moist. For best show, fertilize regularly.
Propagation: From seed, or from tubers. Clean, disinfect, and let tubers dry, storing for at least six weeks prior to replanting.
Note: All parts are poisonous, especially if eaten. Causes mild skin irritation. Deer resistant.

Candleabra Cactus (*Euphorbia ingens triloba*)
Family: Euphorbiacea

Use: The ornamental candelabra growth form makes this a striking specimen for any dry landscape. It has been used as a living fence due to its sharp spines which discourage penetration.
Euphorbias are also grown as houseplants in cooler climates.
Flowers: Small, yellow, clustered in cup-like structures. Fruit is borne on a stalk extending from the flower.
Cultivation: Grows in arid conditions to a height of almost 40', branching from a single trunk. Will grow in hotter, more humid climates as well, but slower growing in those environments. Water sparingly.
Propagation: From seed or cuttings. Take cuttings during dry season, let dry for 10 days prior to planting in cactus mix soil. Seeds stay viable for many years.

Note: All parts of the plant are poisonous. The milky sap is irritating upon contact.

Canna Lily (*Canna x generalis*)
Family: Liliaceae
Alternate Names: Indian Shot (for the hard, round seeds)

Use: Mid-height plant for garden beds, best in massed display. These add a tropical look to any garden, with their large dark leaves and bright flowers. There are over 60 species, but only a few commonly grown.

Flowers: Clear pure bright colors in a wide range, as well as speckled and mottled designs. The leaves are also attractive, large and deep colored in green, bronze, or reddish shades., or striped, as in "Pretoria", left below. The 10' tall *C. iridiflora* carries large flowers on arching stems.

Cultivation: Tolerates almost any good garden soil, full sun to shade. Requires regular watering In cooler climates, cut back to ground prior to cold season. A barrier to spreading roots may be necessary. Cut back old flowering stems for best appearance.

Propagation: By division of the rhizomes, or by seed.

Celosia (*C. argentia " cristata"*)
Family: Amaranthaceae
Alternate Names: Cockscomb, Chinese Wool Flower

Use: Showy, fast-growing annuals with red, gold, yellow, or pink flowers, lasting up to two months. Excellent cut flowers. Raise in greenhouse or conservatory in cool climate. The *Christata* (shown) has fan-shaped, rippled flower heads that resemble coral. Tender leaves are edible. Some medicinal uses. Has been used as a garden ornamental at least since the 1500s.
Flowers: Crested inflorescence, ornately rippled and brilliantly colored.
Cultivation: Requires rich, well-drained soil, with humus and manure added, and kept constantly moist. Needs sun, shelter from wind. Grows to a height of 1 1/2 to 2'.
Propagation: From seed, raised under glass in cooler climates.

Century Plant (*Agave Americana*)
Family: Agavaceae

Use: This native of Mexico will thrive on rocky slopes, steep banks, in sunny gardens or patio pots as an ornamental. The leaves are used to make a beer called pulque, and have some medicinal uses as well.

Flowers: For most of the plant's life, there are no flowers, only a rosette of sword-shaped grey-green leaves. After many years, (30-40, once thought to be up to a century), a flowering spike shoots up several feet from the center of the rosette. Flowers appear on the ends of lateral branches of this spike. Yellow-green clusters bloom, and then the leafy rosette dies, after flowering only once.

Cultivation: Thrives in good potting compost, sunny conditions, ample water during the growth period. Some species of agave yield economically important products, such as liquors and sisal.

Propagation: From offsets near the parent plant. Dry for several days prior to planting.

Coconut Palm (*Cocos nucifera*)
Family: Palmae

Use: The coconut is a palm of many uses. The 60' tree is gracefully ornamental, providing modest shade. Its leaves are used for thatched roofs, baskets, lattice, and fish traps. Its edible fruit contains a liquid used for drink, as a diuretic, and as a sterile solution. The coconut, eaten raw or cooked, has many medicinal uses as well. Oil is used for cooking, margarine, and ice cream. The shells are utensils, and the husk, for matting, padding, and rope fiber. Additionally, the roots provide a dye.

Flowers: Small yellow flowers are borne throughout the year. They are seen on stalks up to 4' long. Mature healthy trees will produce 50-100 coconuts per year.

Cultivation: Light, well-drained, sandy soil in warm climates, especially at low elevations. Very salt-tolerant and hardy. The amount of water and nutrients available influence growth and productivity.

Propagation: The nut has a single seed, which has the distinction of being the largest seed in the plant kingdom, and is capable of floating for long periods before germination. Plant shallowly on its side, 1/3 exposed, transplant at 7-9 months.

Coffee (*Coffea Arabica*)
Family: Rubiaceae

Use: The beans produce a beverage and stimulant used widely for at least 1000 years. There are about 60 species in the wild, about 10 species under cultivation. The most valuable are *Arabica* and *Robusta*.

Flowers: Small, white, star-shaped and fragrant. Flowers give way to red berries.

Cultivation: Thrives in mineral-rich soil, volcanic or freshly-dug woodlands, hilly lands, well-drained, well-irrigated with frequent rains. Prefers wet-hot or hot-temperate climate, altitude of 2000-6500 feet. Plant intermingled with other plants for shelter from excessive sun. Under cultivation, rarely exceeds 7', but *Robusta* can exceed 30'. Requires frequent fertilization.

Propagation: Sow seeds from healthy, long-lived plants in sheltered area, transplant at six months, during rainy season.

Cordyline (*Cordyline terminalis*)
Family: Dracaenaceae
Alternate Names: Ti plant, Good Luck Tree

Use: These lovely plants, native to tropical America and the Pacific, are cultivated for their beautiful leaves, and widely used in landscaping, planted in blocks of one color for maximum impact, or singly as a specimen plant. The green variety is the Hawaiian Ti plant.
Flowers: Small, light pink, appearing at the top leaf cluster. The leaves come in many colors, although the rose-purple is the most popular. There is also a striped variety, in green and white.
Cultivation: Full sun or part shade. Prefers moderate watering. May reach as much as 8' in height. Attains its best color in full sunlight.
Propagation: From seed, stem cuttings, tuber cuttings, or tops of plants.

Crinum augustum
Family: Amaryllidaceae
Alternate names: Swamp Lily, Milk and Wine Lily

Use: African tropical ornamental, with over 130 varieties, used for borders and bedding, or as a water feature plant. Some varieties have medicinal uses in folklore.
Flowers: Produces loose clusters of 20-30 large, lily-like flowers on long, strong stems. Flowers may be red, white, pink, or bicolor red and white, with thin petals.
Cultivation: Flourishes in deep, well-drained to moist soil, in sun or partial shade. May be grown near or in the water for some varieties. Fertilize weekly during growth. The strappy leaves may reach nearly five feet in height in ideal conditions.
Propagation: Grows from large bulbs. Increase by division, lifting and dividing the entire clump. Or, can be grown from seed, which will take 3-4 years to flower.

Crossandra (*Crossandra infundibuliformis*)

Family: Acanthaceae

Alternate Names: Firecracker Flower. The scientific name is from the Greek, meaning fringed antlers.

Use: Tropical shrub originally from India but widely cultivated for its beautiful, long-lasting flowers. Often planted as a low growing (to 3') border. Can be grown as a houseplant or hot-house plant anywhere.

Flowers: Showy salmon-pink to yellow or orange-red flowers in terminal racemes atop a shrub of glossy green leaves. Well-shaped bush with long flowering season.

Cultivation: Light, fertile soil, acid and well drained. Keep moist, cut back after flowering. Does well in a sunny, sheltered, warm place, or indoors, with bright sunlight.

Propagation: By seed, or by cuttings rooted in a peat/sand mixture.

Croton (*Codiaeum variegatum*)
Family: Euphorbiaceae

Use: Hedge or foundation plant, or specimen shrub. Also used as potted plant.

Flowers: Narrow terminal racemes of insignificant white flowers. These are grown for the spectacularly colorful and shapely leaves.

Cultivation: Full sun for maximum color, which ranges from pink, red, yellow, orange, green, to nearly black. Speckled, striped in many variations, also a great deal of variety in leaf shape. Enrich soil with compost and peat moss, fertilize three times a year. Shelter from wind. Water frequently. Grows to 8-10' in warm climate. Does not transplant well. Inspect for spider mites.

Propagation: From greenwood cuttings, leaf bud cuttings, or air layering. Root in sand/peat moss mixture.

Note: There may be more named cultivars of this species than of any other tropical ornamental. Few drawbacks, but there may be some staining from latex sap, and a low level toxicity to mammals.

Datura (*Brugmansia suaveolens*)

Family: Solanaceae
Alternate Names: Angels Trumpets, Moon Flower

Use: In tropical and sub-tropical climates, this fast-growing shrub is an attractive isolated lawn or garden specimen, and is sometimes used in borders. In cooler areas, can be container grown.
Flowers: The single, hanging flowers are the attention-getter for this plant– each may be 10" or longer. They have a musky fragrance, especially in the evening. The narrow, funnel-shaped flowers appear throughout the year. Other cultivars have orange, yellow, or pink flowers, and *B. candida* has double blooms.
Cultivation: Will grow quickly in any frost-free climate. Requires shelter from wind, as it tends to be top-heavy. Prefers fertile, well-drained soil in a sunny or partially shaded spot. The plant will grow to 6', and can be trimmed as a tree. Water generously.
Propagation: By 4-6" cuttings, rooted in a sand-peat mixture.
Note: Datura is attractive to leaf-chewing pests but poisonous to humans due to its content of toxic alkaloids. The plant produces scopolamine, which has medicinal uses, including, in very small quantities, the control of motion sickness.

Desert Rose (*Adenium obesum*)
Family: Apocynaceae
Alternate Names: Impala Lily, Desert Azalea, Sabie Star

Use: Related to the Frangipani but more drought tolerant, this small shrub is evergreen. It grows to 6' in the wild, but is usually smaller in cultivation, often dwarfed as a potted plant.
Flowers: Brilliant blooms of 2" diameter or more, borne in terminal clusters. They range from pink to scarlet, with white and yellow centers. *Adenium* boasts a long flowering season. Does not usually form fruit in cultivation.
Cultivation: Keep warm and well-watered. Prefers partial shade and well-drained sandy soils. Can tolerate drought, and is a good choice for water-conserving gardens. Adapts readily to container culture. It's normal for this plant to lose leaves when it flowers.
Propagation: By seed, cuttings, or simply sticking dried-out branches in moist sand. The seedlings grow rapidly, and will generally flower within the first year.
Note: Sap is poisonous, and used to coat arrowheads in the plant's native range of Africa and the Arabian peninsula.

Dracaena (*D. marginata "Tricolor"*)

Family: Dracaenaceae
Alternate Name: Money Tree, Rainbow Tree, *Pleomele marginata, Dracaena concinna*

Use: Member of the dragon tree family grown for its exotic foliage. Grown as a border plant, or as a specimen. Shows well in massed plantings. Also used as a potted plant in large tubs on patios and balconies. *D. draco* is the stout-trunked Dragon Tree; D.marginata has grey stems and narrow leaves, *D. fragrans* has heavy blue-grey ribbon-like leaves as long as 3'. Grows as a tree or shrub to as high as 20'.

Flowers: Seasonal white fragrant flowers occurring in panicles above the leaves. Some cultivars have blooms of cream or pale green. Fruit is a yellow-orange berry.

Cultivation: Good well-drained soil in full sun to partial shade. Responds well to humid conditions, but let the soil dry between waterings. Prune lightly to promote branching. In cooler climates, overwinter as a house plant. Inspect for spider mites.

Propagation: Cuttings or air layering.

Dwarf Poinciana (*Caesalpinia pulcherrima*)
Family: Caesalpinia
Alternate Names: Barbados Pride, Peacock Flower

Use: Small-tree size shrub, with attractive multihued flowers, excellent along walls, and as hedge border. Can be used to border walkways, but some varieties have thorns. Can be greenhouse grown in termperate climates.

Flowers: The orange-red and yellow flowers with long protruding red stamens appear throughout the year, borne on long stalks in terminal racemes. *C. flava* has yellow flowers. The flowers and leaves somewhat resemble the Poinciana tree.

Cultivation: Native to tropical America, the shrub grows to 13' or more in full sun to part shade. Prefers fertile well-drained soil, and light, regular watering.

Propagation: By tip cuttings placed in a sandy mixture with high humidity, or by seed soaked in warm water prior to sowing.

Fagraea berteroana
Family: Loganiaceae
Alternate Name: Pua-kenikeni, Bishop's Egg

Use: Grown for its fragrant flowers, and also used as an ornamental plant in home and commercial landscaping. The wood is useful for carving and other purposes, and the flowers, for scenting coconut oil.
Flowers: Creamy, yellow-white, small, and highly fragrant, these tubular flowers have long been used in Hawaiian leis.
Cultivation: Native to the South Pacific, this plant thrives in warm, humid climates. Part sun, ample water in well-drained soil for best growth. Can reach a height of 30', but most often seen as a medium-sized shrub.
Propagation: From seed, cuttings, or air layering. Often starts as an epiphyte or strangler.
Note: Mildly toxic.

Ferns (*Pteridophytes*)

Use: Ferns are cultivated for their attractive appearance, and are Used in groundcovers, rock gardens in shaded areas, hanging baskets, against walls, and as container plants. They are an importent part of the forest understory in the wild.

Flowers: Ferns do not produce flowers nor do they bear seeds or fruit.

Cultivation: Most species require part shade, with dappled sunlight. Water at least weekly, and fertilize occasionally for best results. Tropical ferns rely on humidity for sustenance, so most prefer to be grown out of doors.

Propagation: Many ferns have microscopic spores protected by the sori, usually found on the underside of the fronds. Some types, called Mother ferns, produce bulbils, or pups. Most ferns can be propagated by division, and it is possible to reproduce ferns from the spores, although this process requires patience.

Note: The selection of ferns shown here for illustration were all found in the general area of Puerto Vallarta, a tropical coastal Mexican city.

Ficus (*Ficus benjamina*)
Family: Moraceae
Alternate Names: Weeping Fig, Benjamin's Fig, Banyan

Use: The gigantic *ficus* trees found in Mexico are related to the fiddleleaf fig and to the weeping banyan, as well as to the small shrub thought to have provided leaves for Adam and Eve's clothing. They are seen as huge specimens with aerial roots and strangling branches, and also as tightly trimmed topiaries shaped imaginatively to amuse the tourists. Over 600 species of *ficus* exist. *F. benjamina* is planted as an ornamental or street tree due to its graceful form.
Flowers: None. Fruit is an orange, pink, red, or purple round form, found in pairs.
Cultivation: In the wild in tropical or subtropical climates, can reach massive size and height. Cultivated and trimmed as a yard specimen, needs full sun and regular watering. Avoid planting close to structures, as its massive underground root system can be damaging. Has a milky sap.
Propagation: From aerial roots or branches that touch the ground and establish new individual trees. Can be propagated by cuttings or air layering. Considered invasive in some places.

Ficus elastica
Family: Moraceae
Alternate Names: Indian Rubber Tree, Rubber Plant

Use: Native to South America, it is cultivated widely for rubber production from its latex sap. Also grown as a landscape specimen in large gardens or public spaces, and in northern climates, as a houseplant. Mesoamericans historically utilized the latex rubber for balls for games, temporary shoes, padding for tool handles, and for making water-resistant cloth.

Flowers: Not highly colorful or fragrant, the flowers are only pollinated by a particular species of fig wasp, and attract primarily those wasps in a highly evolved relationship.

Cultivation: As a houseplant, requires bright light but not direct sun, and will lose leaves if environment changes too much. Outdoors, it grows rapidly to over 100', and where it receives less light, it produces larger leaves. Drought tolerant, but does not do well if in flooded conditions. Commercially grown in India and elsewhere in plantation rows. Responds well to twice-annual feeding.

Propagation: By seed, which are contained in fruit where the fig wasp species is present.

Fig (*Ficus carica*)
Family: Moraceae
Alternate Name: *F. lyrata,* Fiddle-leaf Fig, Banjo Fig

Use: This member of the large *Ficus* family bears edible fruit, and grows to well over 40' in the tropics. The tree is characterized by flat planes of the large, glossy, fiddle-shaped leaves. Its distinctive shape lends itself to landscape uses. Can be container grown as well.
Flowers: The flowers are borne inside a small hollow receptacle, and therefore are not visible to the passerby. Fertilization is carried out by a tiny wasp.
Cultivation: Full sun. Thrives in evenly moist soil, with biweekly feeding during periods of active growth.
Propagation: New plants can be started from cuttings, or by air layering. Protect roots near the surface.

Fishtail Palm (*Caryota rumphiana*)
Family: Arecaceae
Alternate Names: *Caryota*, Wine Palm

Use: Attractive tropical garden specimen, with tall straight trunk and drooping, forked fronds.
Flowers: Produced at maturity, the flowers of some types may feature drooping panicles of up to 20' in length. The plant is primarily grown for its striking foliage, which resembles the forked tails of ornamental fish.
Cultivation: Requires heat and humidity. Very fast growing. Can tolerate some expended periods of low light, or permanent part shade conditions. Highly salt tolerant. May reach heights in excess of 25'.
Propagation: From seed or from offsets. In the wild, the suckering varieties will develop into a dense thicket.
Note: Pulp of the fruit contains oxalic acid crystals that sting. Seed kernels are edible.

Flame Vine (*Pyrostegia venusta*)
Family: Bignoniaceae
Alternate Names: Llamarada, Orange-Flowered Stephanotis, (also called Honeysuckle in error)

Use: Spectacular, fast-growing climbing vine that will rapidly cover a wall, arbor, or trellis. Originating in Brazil, this popular tropical vine is a favorite of hummingbirds. Can be container grown as well.
Flowers: Produces curtains of dense clusters of brilliant orange, tubular 3" flowers from Fall through Spring. Each cluster holds 15-20 of the flowers, which are followed by 1' long dry capsules.
Cultivation: A woody evergreen climbing vine, it tolerates a variety of soil types, needs full sun and moderate water. Prune hard after flowering. Can be aggressive to the point of smothering a host shrub or tree.
Propagation: From semi-hard cuttings in summer.

Foxtail Palm (*Wodyetia bifurcata*)
Family: Areceae

Use: Unique, full-appearing frond shape plus neat appearance make this medium-sized palm a select specimen for landscapes.
Flowers: Creamy green bloom, borne from a green inflorescence borne on a thick stem, followed by oval (1.25" x 2.25") fruits that turn an attractive orange-red, with a single seed.
Cultivation: This fast-growing palm is salt, wind, heat, and cold tolerant. It has been found growing in open woodlands, in stands along creeks, and in rainforests, and does well in many types of soil. It is self-cleaning. Fertilize heavily, water regularly. Grows to 30' in sun or shade.
Propagation: From seed, which germinates in 2-3 months. Plant in well drained sandy medium, keep moist.

Ginger (*Zingiber*)
Family: Zingiberaceae

Use: Many gingers are staples of the floral industry, with their flashy, long-lasting blooms. True ginger (*Zingiber officinale*) has been known for the culinary and medicinal applications of the ginger root since at least 500BC (lower left). All of the gingers have aromatic rhizomes.

Flowers: All of the more than 50 species produce their flowers on separate stems from the leaves. The "torch" produced by the torch ginger grows singly from the ground on a long stalk, and consists of large flower heads of waxy overlapping bracts with small flowers.

Cultivation: Gingers grow to full height of about 8', in one season. Prefers deep, moist soil, can be grown as a houseplant as well. Those with narrow leaves require full sun; those with wider leaves can tolerate partial shade. Feed biweekly during growing season.

Propagation: Easily grown from seed, or propagate by rhizome division.

Guava (*Psidium guajava*)
Family: Myrtaceae
Alternate Name: Guabaya

Use: This attractive small (to 33') tree, a native of Mexico or Central America, is grown widely in tropical and subtropical climates for its fruit, cultivated in many varieties, and eaten raw or cooked. The wood is used in carpentry and for fuel, the leaves, for dye, the bark, for tannin, and the "wood flowers" caused by parasites, sold as ornamental curiosities.
Flowers: Small, white, and lightly fragrant, followed by round or pear shaped fuits to 4" long.
Cultivation: Frost sensitive as well as to intense heat, but will thrive in both humid and dry climates, as well as in a variety of soils. Prefers good drainage but will grow in wetter land than other fruit trees. Somewhat salt tolerant. Prune to eliminate suckers.
Propagation: By seeds, which remain viable for several months. Boiling or soaking hastens germination. Cuttings, air-layering, and grafting are also used as propagation methods commercially.

Gumbo Limbo (*Bursera simaruba*)

Family: Burseraceae
Alternate Names: *Papillilio*, Tourist Tree, West Indian Birch

Use: Attractive specimen due to its shiny red peeling bark, used for street plantings, also used as a living fence, as it will form thickets. Produces a turpentine-smelling resin with many uses, and the soft, light, wood is easily carved but brittle. Birds eat the small fruits. In Haiti, the trunk is used to make drums. Bark has medicinal uses.

Flowers: Inconspicuous small flowers of 3-5 greenish petals, followed by slow-maturing dark red fruits. Other members of this family produce the incenses myrrh and frankincense.

Cultivation: Grows in open hammock woodlands from Florida south to Venezuela. Loses leaves in dry season, but is drought tolerant and salt tolerant. The trunk expands to store water during the rainy season. Plant in full sun to part shade Grows 25-50' tall, and the spreading canopy of featherlike leaves on randomly angled branches may reach 35-50' in width.

Propagation: Easily grown from seed, or simply stick a branch in moist soil. Fallen trees send up suckers.

Note: The name "Tourist Tree", or in Mexico, "Gringo Tree" is due to the peeling red bark, which looks as if the tree had suffered a sunburn.

Guzmania (*Guzmania lingulata minor*)
Family: Bomeliaceae
Alternate Names: Red Cockade, Golden Cockade

Use: These hardy bromeliads are grown primarily for their long-lasting (as long as 3 months), decorative blooms. Leaf rosettes are about 9" long. Excellent house plants if misted twice weekly.

Flowers: The small, tubular, whitish-yellow flowers are surrounded by colorful bracts of orange red, gold, or green. In hot-house conditions, blooms can be forced year-round. Some varieties, such as "Lady Alice" have white-edged leaves with yellow flowers atop the central stalk.

Cultivation: Garden soil, or in pots of open, rubble-filled compost. Prefers filtered light, constant humidity. Leaf vases should be filled with water all through the summer and flushed occasionally.

Propagation: Offshoots are freely produced. New plants need to be separated from the mother plant for them to flower. Dead flowers can be cut off and potted for new plant development.

Halfmens (*Pachypodium*)
Family: Apocynaceae

Use: As landscape specimen, or indoors, as a potted plant similar to a cactus.

Flowers: After the rains, leaves appear, then pale yellow, pink, or white flowers appear in the center of the leaves. The flowers are much like *Plumeria* in appearance. The plant may be thirty years old before flowering.

Cultivation: For half the year, this desert shrub is barren of leaves and flowers, appearing as a cactus. The swollen upright branches are covered with thorns. It requires a warm to hot climate, little water. If potted, it should be taken indoors in cool seasons. Avoid excess moisture by using clay pots. Slow growing, to a maximum height of around 15'.

Propagation: From seed.

Note: The name is from their appearance, half like men, nodding and talking together.

Heliconia (*Heliconia caribeae*)
Family: Musaceae
Alternate Names: Parrot Flower, Lobster Claw

Use: Spectacularly beautiful group of plants from the tropical family which includes bananas, travelers' palm, cannas, and bird of paradise. Flowers used in showy tropical arrangements. The plants are marvelous landscape showpieces in warm, humid climates. A few are grown for foliage. Reaches 8' in height.

Flowers: The tiny flowers are hidden inside dazzling colored bracts in green, yellow, pink, red, and orange, often combined. The showy bracts are over 12" long. Some varieties, such as the fishpole heliconia (above, left), have bract clusters that dangle, others, like golden heliconia, are upright.

Cultivation: Native to South and Central America, as well as some South Pacific islands, they also are grown in warm greenhouses or conservatories further north. They thrive in rich compost with generous water and filtered sun. Found in humid lowland tropics below 1500 feet, and in rain forest habitats.

Propagation: Hummingbirds and nectar feeding bats are known pollinators.

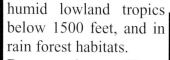
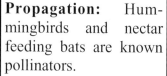

Huernia (*H. hystrix, H. schneideriana*)
Family: Asclepiadaceae

Use: Generally, rock gardens or subtropical gardens. Especially effective in raised containers to allow the stems to hang down over the edge. In cooler climates, often used as a hothouse plant or indoor potted plant.
Flowers: The small flowers have a bell shape with a five-pointed outer edge, and are brownish red, with darker, almost black, centers.
Cultivation: This easy-care succulent does well in ordinary garden soil in full sun or indoors in bright light. Moderate water needs, but requires more water during bloom season.
Propagation: From cuttings, callused prior to planting.
Note: When this plant was named by Robert Brown, a mis-spelling occurred, so that it is not clear that the name is intended to honor the seventeenth century collector of plants from the Cape of Good Hope, Justin Heurnius.

Hibiscus (*H.rosa-sinensis*)
Family: Malvaceae
Alternate Names: Rose of China, Rose Mallow, Rose of Sharon, Shrub Althaceae, Rosella

Use: Magnificent evergreen shrub to 6', for garden, balcony, terrace, or in large pots. Produces black dye, and has some medicinal uses. In Jamaica, its petals are used to polish shoes.

Flowers: Free-flowering. Both single and double (*odorata*) varieties are common. *H.schizopetalus,* from East Africa, (lower right) has slashed petals. Blooms may reach 6-10", but when cut, last only a day.

Cultivation: Well fertilized loamy soil with sand and peat addition. Frequent, generous watering, full sun. Do not let dry out. May require staking. Inspect for spider mites.

Propagation: From 4" cuttings set in a mixture of peat and sand.

Huanacaxtle
Family: Legumeinasae
Alternate Name: *Parota*

Use: Farmers plant the *Parota* outside of town to provide shade for their cattle, and use the high-protein leaves in cattle feed as well. In town, the graceful, spreading trees provide shade in parks, and the termite-resistant wood is widely used for doors, window frames, and furniture.
Cultivation: The fast growing *Parota* is found in a variety of soils, in full sun. It can reach more than 90', with a trunk of 10' or more in diameter. The canopy is smoothly rounded, and where there is space, the tree makes an attractive landscape specimen.
Propagation: From seed.

Ixora (*I. chenensis*)
Family: Rubiaceae

Alternate Names: Jungle Geranium, Mexican Geranium, Flame of the Woods, Jungle Flame

Use: Attractive landscape plant, often used in hedges and borders, flowering throughout the year.

Flowers: Striking blooms in shades of pink, orange, red, yellow, and white, with reds and oranges most common. Flowers have some medicinal uses.

Cultivation: This small shrub, about 3-6 ' high, is related to the coffee plant. It grows in any soil, in sun or part shade, along the streets, or near the sea, but prefers sandy soil enriched with leaf-mould. Must be pruned to look neat, and fertilized to encourage rapid growth. Can be grown indoors. Only moderate drought tolerance. Best suited for gardens where the humidity is high year-round.

Propagation: Some varieties can be propagated by cuttings, most can be propagated by seed.

Jatropha
Family: Euphorbiaceae
Alternate Name: Coral Plant

Use: Interesting garden specimen, particularly the *J.podagrica*, for its swollen stem, which makes the plant resemble a small baobab tree. *Jatropha curcas* is a multipurpose crop grown in India and elsewhere to prevent soil degradation, and is used for bio diesel, for the production of soap and tannin. Oil is extracted from the seed, dye and wax are produced from the bark and roots. Its juice has medicinal properties as well.

Flowers: The coral-red flowers of the variety shown are borne almost year-round in tropical gardens. Other varieties have purple flowers (*J. gossypifolia*).

Cultivation: *Jatropha* is a spreading, soft-wooded shrub from the Caribbean, and the tallest varieties grow to approximately 10' tall. It does well in full sun.

Propagation: From seed, or from cuttings, which root easily.

Jellybean Plant (*Sedum rubotinctum "Aurora"*)
Family: Crassulaceae
Alternate Name: Pink Jellybeans, Christmas Cheer, Pork and Beans

Use: This Mexican native is an excellent ground cover for hot, dry areas, can also be container grown, or used as an indoor plant, providing it receives at least four hours daily of direct sunlight. Plant in its permanent place, as it loses leaves when moved.
Flowers: Yellow flowers borne amidst the leaves.
Cultivation: This hardy plant tolerates a wide variety of soils, and part shade to full sun. The more light it receives, the more brilliant its color. One of the easiest to grow small succulents, it is drought tolerant, but not at all frost tolerant. Water weekly in hot weather, less often as the weather cools, cutting back to once a month in the coolest season. For the best appearance, provide full sun and good drainage. Check for mealy bugs.
Propagation: New plants spring up from leaves, or "beans", which drop off.

Kalanchoe (*K. brasiliensys, pinnata*)
Family: Crassulaceae
Alternate Names: Air Plant, *Iverea pinnata*, Mexican Loveplant, Miracle Leaf

Use: This large group of succulent plants, native to Mexico, (some 150-200 species), serve as accent, border, or woodland plants, and are frequently grown as houseplants, for the flowers and the attractive downy leaves of many species. These plants have a longstanding history of medicinal use, particularly among indigenous peoples of South America, who use the leaves and leaf juices to treat a wide variety of ailments.
Flowers: Borne in dense clusters high above the leaves. In the variety pictured lower left, the flowers are tubular and pendulous. Colors range from the lavender and greens shown to red, orange, white, and pinks. The flowers are long-lasting, making a good potted plant gift.
Cultivation: Full sun to part shade. Water thoroughly, let dry to touch. Requires well draining porous soil of half loam, and equal parts sand and broken brick. Fertilize every three weeks while in bloom.
Propagation: From seed, or from tip cuttings, which root within two weeks. Plantlets grow on the leaf tips and will readily develop into new plants. Many species produce these plantlets quite prolifically.

King Palm (*Archontophoenix alexandrae*)
Family: Arecaceae
Alternate Name: Alexander Palm

Use: Landscape specimen or streetscape planting in coastal areas. Also used to stabilize soil erosion. This tall (to 80'), stately tree has a smooth greenish gray ringed trunk with leaf scars and an enlarged base, and silvery green leaves 8' to 14' long.

Flowers: Creamy yellow-white, coming from a flowerstalk that emerges from below the crownshaft. Fruit turns to an attractive red when mature, and each contains a single seed.

Cultivation: Full sun to light shade, shelter from high winds. Needs regular, moderate watering, as it naturally occurs in swampy areas or near streams. Grows rapidly, 1-2' per year. Tolerates a variety of soil types, but is not particularly salt tolerant. Due to its natural affinity for water, this palm can tolerate moist soil. Leaves are self-cleaning.

Propagation: From seed, which germinates easily, within 3 months. Best results if planted from container.

Lantana (*L. camera*)

Family: Verbenaceae
Alternate Names: Red Sage, Shrub Verbena, Wild Sage, Yellow Sage

Use: Popular landscape plant for its bright flowers and tolerant disposition, growing in many poor or dry soils. Can be container grown as well. This vigorous shrub attracts butterflies and hummingbirds. The stalks are used in making paper pulp, the bark produces an astringent lotion, and the leaves have medicinal uses as well. The plant also concentrates nitrogen in soil.
Flowers: The small single or bicolor clusters occur in many colors and combinations, including white, lavender, pink, red and orange, changing over time. The flowers are followed by fruits which are poisonous.
Cultivation: Full sun to part shade, moist but not wet soil. However, will tolerate dry and poor soil. A hardy shrub, it grows to about 6' in height, but there are dwarf varieties available.
Note: The leaves are poisonous to cattle, and because of the devastating impact of this hard-to eradicate weed, it is perhaps the most studied weed in the world. In the wild, it is invasive and difficult to eradicate, regenerating from basal shoots even after fire. In the garden, however, it is prized for its attractiveness to birds and butterflies.

Licuala Grandis (l.), Licuala Spinosa (r.)

Family: Arecaceae
Alternate Names: *L. spinosa*; Mangrove Fan Palm, Good Luck Palm, Spiny Linuala
L. grandis; Ruffled Fan Palm, Vanuata Fan Palm, Ruffled Lantan Palm

Use: These slow-growing palms are excellent for indoor display. In outdoor cultivation, they grow to 10-15'. *L. spinosa* has multiple clumps and the appearance of a dense bush. Each leaf section has a squared-off end, creating an attractive and unusual appearance. *L. grandis* leaves are circular, undivided and regularly pleated, about 22" in diameter.
Flowers: In both, the flower stalk emerges from among the leaves, and the flowers are followed by red marble-sized berries.
Cultivation: Full sun for *L. spinosa*, light shade for *L.grandis*. High humidity and lots of water are the elements of an ideal *Licuala* environment. *L. grandis* prefers sandy loam soil or limestone, and protection from wind. *L. spinosa* can withstand some frost, as well as poorly draining soil, but has low salt tolerance.
Propagation: From seed, and in the case of *L.spinosa,* from clumps.

MacArthur Palm (*Ptychosperma macarthurii*)
Family: Arecaceae

Use: Widely distributed ornamental in tropical and subtropical areas. Looks wonderful under a canopy of tall trees. When young, makes a lovely container plant, or provides a stunning background landscape plant against walls. The greenish grey trunk and leaves which look as if they were cut with shears are in contrast to the otherwise similar Butterfly Palm.

Flowers: Inconspicuous, borne on an inflorescence. Flowers are greenish yellow to cream, and are followed by green fruit which ripens to red, borne in long pendulous clusters.

Cultivation: This elegant slender clumping feather palm grows to 30' with a 10' spread and has leaves of 12" or longer. Will tolerate full sun, but prefers part shade, tolerates a variety of well drained humus rich soils. Needs adequate water in dry season, and regular feeding. Not particularly drought or cold tolerant. Prune sparingly to avoid damaging lower trunk or roots. Naturally occurs as an understory plant in tropical forests. Cold-sensitive. Protect from wind.

Propagation: From seed, which germinates in 2-3 months. Transplant immediately after germination or after 1-4 leaves have formed.

Mandevilla (*Mandevilla*)
Family: Apocynaceae
Alternate Names: Mexican Love Vine, Chilean Jasmine, *Dipladenia*

Use: One of the favorite, most ubiquitous ground covers and vines in commercial and home landscaping. Also can be grown in pots and indoors, as long as the large root system is accommodated.

Flowers: These charming yellow flowers are everywhere– climbing over patios and roofs, around street sculpture bases. The flowers are large, abundant, and have a long flowering season. There are soft pink varieties (*M. amoena,* Alice du Pont, below, right), a mauve pink variety, and several white to pale pink varieties as well.

Cultivation: Does well in many different types of soil, as long as there is regular water and occasional applications of liquid fertilizer. It's best to keep the soil moist during flowering season, and prune vigorously to contain the spread. Full sun or partial shade.

Propagation: From seed, or from hardwood cuttings, which sprout in about four weeks.

Mango (*Mangifera indica*)
Family: Anacardiaceae
Alternate Names: Peach of the Tropics, King of Fruits

Use: This member of the cashew (sumac) family is a handsome upright landscape tree which produces an abundance of edible fruit.
Flowers: At the branch tips, curved upward-pointing panicles of tiny pink flowers appear, each with four or five petals. These are followed by the heavy, slow-ripening 6" fruits.
Cultivation: Found throughout the tropical and sub-tropical climates, the mango requires a warm climate and abundant moisture, but can tolerate a dry season of several months. Grows best in full sun. It may reach 100' in Southeast Asia, but generally less than that in Mexico.
Propagation: Ordinary seedlings bear well, although the fruit may be somewhat fibrous. The improved varieties are grown from cuttings and are free of fiber.
Note: The leaves have a turpentine odor when crushed.

Manilla Palm (*Veitchia merrillii*)

Family: Arecaceae
Alternative Name: Christmas Palm

Use: The Manilla palm is a small, widely planted ornamental, primarily known for its showy clusters of 1" red fruits ripening about Christmastime. Used as a container palm, as well, due to its smaller size and neat appearance.

Flowers: The white or yellow small flowers form below the crownshaft, and are followed by clusters of green fruits which turn crimson red.

Cultivation: Does well in sun or part shade, good light and dry soil. Has a stocky single trunk with stiffly arched leaves. The leaves are feathery, each with many upswept leaflets that are widest in the middle. Grows to 25', with a 10" diameter trunk, although often smaller. Frost sensitive, as is common with South Pacific natives. A bit vulnerable to yellowing disease.

Propagation: By seed.

Note: Seeds are sometimes used as a substitute for betelnut.

Maritime Pine (*Pino maritimo*)
Family: Casuarinaceae or pinaceae

Use: Used for windbreaks and as an ornamental in large gardens. The bark extract has medicinal applications in people and animals. The dense, strong timber is used for lumber and in the manufacture of particle board and fiber board. The inner bark, young cones, and a tea made from the needles are all sources of nutritious survival food. Maritime pines have also been used for stabilization of coastal areas threatened by sand dunes, and, experimentally, in the creation of carbon sinks.

Cultivation: The maritime pine originated in Australia, and is one of about 40 varieties of pine found in Mexico. It prefers acidic soil with good drainage. It is hardy, both drought and salt tolerant.

Propagation: By seed. At maturity, the cones open and release the seeds, which are distributed by wind and animals.

Mexican Fan Palm (*Washingtonia robusta*)
Family: Palmae/Arecaceae

Use: Streetscapes, formal plantings near tall buildings, parks, playgrounds. Can be container grown when young. A striking, tropical look in a hardy palm. Native to northern Mexico, this palm is very common in California and Arizona as well. Leaves used in building palapas.
Flowers: The light beige branched inflorescences hold clusters of small white flowers, which are followed by black berries about 1/2" in size.
Cultivation: This salt resistant, drought resistant and cold hardy palm will survive in many conditions, but does and looks better with sun to part sun, and better quality soil. It retains a "hula skirt" of dead leaves, giving a distinctive look to this tall (over 100') palm. The trunk slims as it rises, and the crown of fan-shaped leaves forms a oval "head" high in the air. This palm is quite fast-growing, at a gain in height of about 2' per year. A low-maintenance tree, it can be pruned for a cleaner appearance.
Propagation: From seed, which germinates readily, within two months.

Mexican Oil Palm (*Attalea cohune*)
Family: Aracaeae
 Alternate Names: Cohune Palm, American Oil Palm, Corozo Palm, Majestic Palm, Rain Tree.

Use: Large (to 90'), dramatic specimen plant for parks, public gardens. Leaves used for thatch, trunk for construction, seeds for fodder and oil, heart is considered a delicacy, palm wine is made from sap.
Flowers: Large clusters to 5' of cream colored flowers set among the leaves, followed by brown-yellow fruits on long drooping stalks.
Cultivation: A Mexican native, grows in tropical and subtropical areas in full sun, adapts to many soils, but needs good drainage. Responds well to regular feeding. Pest and disease resistant. Trunk can reach 50' and 2' in diameter, with erect leaves up to 33' long, arching at midpoint. Solitary trunk covered with old leaf bases may grow underground for many years.
Propagation: From seed, with 1-3 seeds contained in each fruit.

Morinda Noni (*Morinda citrofolia*)
Family: Rubinaceae
Alternate Names: Painkiller Tree, Indian Mulberry

Use: This member of the coffee family is known primarily for its medicinal and cosmetic uses throughout the tropics. All parts of the plant are useful—even the trunk is used for tools and firewood. The leaves and mashed fruit are said to be excellent poultices for wounds, and a medicinal drink is made from the fruit. Folklore considers this plant a cure-all, and the odorous fruit is eaten in times of famine. The roots produce a yellow dye, used for coloring tapa, or bark cloth.

Flowers: Small white, 5 lobed flowers. When the fruits appear with their markings, each section represents the place where a flower was once attached.

Cultivation: *Noni* tolerates most soils and moisture conditions, as well as salt and wind. It grows best in full sun to an altitude of 2600', reaching a height of 10-20'.

Propagation: From seed, which remain viable for long periods.

Morning Glory Tree (*Ipomoea arborescens*)
Family: Convolvulaceae
Alternate Names: Dawn Flower, Moonflower, Palo Blanco

Use: The familiar morning glory vine or ground cover is in Mexico sometimes found as a deciduous shrub or tree with a woody trunk and branches, growing to 20'. It is seen in roadside plantings, small yards, and in the wild. The large flowers are quite attractive. There is a grand specimen of this unusual tree in Kapiolani Park in Hawaii, a similar climate.

Flowers: Most often seen in white with red centers, there are also shrub varieties with flowers in blues and purples (lower right, upper left), which open early in the morning, and fade later in the day. The Japanese varieties of *Ipomoea* show many more colors, including mauve, pink, and a beige-edged white.

Cultivation: Will grow in nearly any climate, but prefers a cool, not too humid environment. Does best in full sun, with frequent generous watering. Prefers fertile soil, consisting of loam, sand, and peat.

Propagation: From seed. Annual in temperate zones.

Mussaenda (*M.* frondosa)
Family: Rubiaceae
Alternate Names: Flag Bush, Buddha's Lamp, Handkerchief Plant

Use: Spreading shrub to 10', with spectacular display that lasts nearly year-round. Used in home and commercial gardens, parks, and streetscapes. Often grown standing alone, as a specimen plant. This is a member of the coffee family, and its flowers are said to have medicinal uses.
Flowers: The actual flowers are small and insignificant, but the large pink or white bracts provide a lovely display of bloom resembling poinsettias and lasting for months. Flowers continuously throughout the year. *M. philippica*, a Philippine native, has white flowers. *M. erythrophylla*, a Ghanian native, has bright red petal-like calyx lobes with white flowers.
Cultivation: Prefers a light soil and ample water in full sun to part shade. Cold sensitive below 59 degrees F. Prune regularly to force more flowers, and to maintain size and shape.
Propagation: From hardwood cuttings, kept warm and moist.

Nopal (*Opuntia ficus-indica*)
Family: Cactaceae
Alternate Names: Prickly Pear Cactus, Barbary Fig

Use: An important symbol of Mexico, this cactus appears on the national flag. Probably cultivated for at least 5000 years, it has a long history of food and medicinal uses. The stems are eaten as a vegetable, the fruits are eaten raw or fermented, and the plant provides an important source of vitamins, protein, and minerals. It has been used as a medicine for diabetes, liver disease, easy bruising, and hangovers. The plant is a traditional property line marking hedge. In addition, the Nopal provides topsoil protection, cattle feed, dye, rubber, and an anticorrosive.

Flowers: The red, pink, yellow, or orange flowers open at dawn or noon. The fruit is the prickly pear.

Cultivation: This native tree or shrub is found in dry conditions, growing in full sun to 15' in height.

Propagation: By offshoots, or simply by cutting a *penca* (leaf, or pad), and putting it in the dirt, where it will root rapidly.

Norfolk Island Pine (*Araucaria excelsa heterophylla*)
Family: Araucariaceae
Alternate Name: Star Pine

Use: Majestic conifers, naturally found only in the Southern hemisphere. Native to Norfolk Island and also found in coastal north-east Australia. This decorative tree was widely used in beachfront plantings since it is normally salt-resistant. Valuable in preventing seaside erosion. Also used for timber, decorated as Christmas trees, and used widely as houseplants. In the wild, grows to 200', although lightening often limits its growth.
Cultivation: Grows rapidly in full sun, will survive in partial sun. Let the soil dry to 2" deep between waterings, feed twice annually with organic fertilizer.
Propagation: From seed. Male and female cones are found on separate plants. Cuttings may be taken from terminal shoots.
Note: The blue and black birds in center photo are San Blas Jays.

Oleander (*Nerium oleander*)
Family: Apocynaceae
Alternate Name: Rosebay

Use: Widely used as garden specimens, in roadside plantings, and as an ornamental shrubs for entrances, hedges, and the like. Potted plant in cooler climates. Can grow to 20'. Some types have variegated foliage, some are highly fragrant.
Flowers: In dense clusters at the tips of new wood, single or double, in white, pink, red, purple, salmon, and yellow varieties. Has a long flowering season.
Cultivation: Grows well in ordinary garden soil, in full sun or part shade, and with generous water. Fertilize lightly every week, prune for shape. Stake for support. Red variety is somewhat frost tolerant. Salt tolerant and drought resistant.
Propagation: Plant shoots in a standard sand/peat mixture, water well. Can be grown from seed also.
Note: All parts of the plant are poisonous, and fatal if ingested. Avoid smoke from burning plants.

Orchid Tree (*Bauhinia variegata*)
Family: Caesalpiniaceae
Alternate Names: St. Thomas Tree, Butterfly Flower

Use: Beautiful specimen tree that grows to 30'. The shrub varieties are sometimes used as a hedge.
Flowers: Lengthy bloom season of elegant flowers that resemble orchids, in lavender or white. Other varieties bear red, pink, or yellow blossoms. All have the twin-lobed leaves. The flowers are followed by 6-9" woody pods. *B. variegata*, shown, has five fertile stamens.
Cultivation: Requires moderate water year-round. Full sun or light shade preferred. Hardy to 25 degrees F. Tolerates humid tropical and drier subtropical climates. Grow in rich, well-drained soil, and prune lightly to shorten the branches after flowering, both to maintain shape and to prevent litter from the production of many seed pods.
Propagation: From scarified seed sown direct.
Note: Because of the unusual twin-lobed leaves, this plant was named for two 16th-century brothers, perhaps twins, who were botanists.

Orchids
Family: Orchidaceae

Use: Grown world-wide for their beauty, a variety of orchids also thrive in Mexico, some at sea level, but most at altitude. Of the world's 25-30,000 species, many of the smaller ones are especially adapted to conditions in Mexico.

Flowers: Orchids have a modified petal (labellum), referred to as the lip, which is also the source of scent in those few types which are scented. Orchids come in most colors, with light purple, yellow, and white common. Orchids have a long flowering season, and cut flowers last a long while as well.

Cultivation: Orchids are epiphytic, living on trees or rocks in the tropics, and in soil in temperate zones. Most need high humidity and frequent watering during active growth, Good ventilation of the roots is important, as is good light but not full sun. Do not fertilize during dormancy.

Propagation: Male and female parts are fused on the column of the flowers.

Note: Vanilla (vine at lower right) produces pod-like fruits that are the source of vanilla flavoring.

Organ Pipe Cactus (*Stenocereus thurberi*)
Family: Cactaceae

Use: Landscape specimen which stores water, and bears edible fruit. The fruit is made into jams and beverages.
Flowers: Night-blooming, small (2 1/2") lavender-white flowers borne laterally near the apex of the stems.
Cultivation: Grows at about 1000' in Mexico and Arizona. Native to the Sonoran desert, it is the second tallest cactus, growing to about 23'. Forms a cluster of 5-20 slender branches that curve upward from the ground. Each 6" diameter branch has 12-17 rounded ribs. These stems rarely branch, but grow from the tip each year.
Propagation: From seed. Needs shade during seedling stage. Usually pollinated by bats.

Oyster Plant (*Rhoeo discolor*)

Family: Commelinaceae
Alternate Names: Boat Lily, Moses in the Bulrushes, Liver Leaf

Use: Ornamental long-flowering plant for shady areas, often used in container gardening. This plant, native to Mexico, flowers all year long. Has folkloric medicinal uses. Often used for groundcover, *Rhoeo discolor* can be invasive.

Flowers: Small white flowers appear nested at the base of the upper leaves, surrounded by light purple bracts. The undersides of the leaves are purple as well.

Cultivation: Needs some shade and frequent watering, occasional fertilizer. Does well in any good potting or garden soil.

Propagation: By seed, cuttings, or division of the mother plant.

Papaya (*Carica papaya*)
Family: Caricaceae

Use: Tropical shrub and small tree, with edible fruit which is also processed for juice. The milky sap from the stem and unripe fruit contains a substance, Papain, which is used as a tenderizer for meat.

Flowers: Fragrant, 5-lobed flowers clustered on long, many-flowered stalks. Some varieties have pistillate flowers which are tubular, with separate petals, borne 1 to 3 on separate stalks. Fruits are similar to melons, yellow green to orange, 4" or longer, with sweet, white to orange flesh and many dark seeds. Papaya, first described in 1526 by the Spaniards, is a reliable food source, bearing fruit for as long as 20 years.

Cultivation: This small tree grows to 20' and about 6" in diameter. With sufficient moisture, the tree can reach 30'. These southern Mexico native plants grow rapidly in a warm climate with abundant moisture or irrigation. Space plants 8-10' apart in well-drained soil. Protect from wind and frost, plant in full sun. Regular fertilization will improve fruiting. Thin overcrowded fruits when young. The papaya will bear fruit in 8-10 months.

Propagation: Grafting and rooting of cuttings are effective, but labor intensive. The seed germinates readily, and is viable for up to 3 years under cool dry conditions. Hand pollinate for best results.

Papyrus (*Cyperus papyrus*)
Family: Cyperaceae
Alternate Name: Egyptian Paper Plant

Use: Pith used to make paper from the 3rd millennium BC until the 3rd century AD. Thought to be the bulrushes that sheltered the infant Moses. Used in indoor containers, by indoor pools, as an aquatic beside ponds and lakes, as landscape specimen bordering water features.
Flowers: Yellow to greenish brown "mop-heads" with red in the stamens
Cultivation: Well drained soil, with charcoal and bonemeal added. Needs constant moisture. Has a long principal root stalk, with its roots descending vertically. Triangular emerald green bare stems topped by large round straw yellow umbrellas. Can reach a height of 19 feet and a stem diameter of 4 inches.
Propagation: By immersing the head on a 6" stalk upside down in water. Will root in 10-14 days.

Philodendrons
Family: Araceae

Use: Indoors, in hanging baskets, ground cover, borders, against walls, as climbers., as in the tropics, where they cover trees.
Cultivation: Provide regular ample water and regular feeding, part shade. Not at all drought tolerant. Climbers need support.
Dust the leaves to enable maximum use of available light. Variegated varieties need more light than the solid green types.
Propagation: From stem cuttings.

Pygmy Date Palm (*Phoenix roebelinii*)
Family: Arecaceae
Alternate Name: Phoenix *humilis* Royale

Use: The graceful, arching fronds of up to 8' on a relatively small palm, to about 10', make this an ideal courtyard plant, or specimen plant for a small garden. This palm is sometimes used in interiors due to its tolerance of lower light levels.

Flowers: Inconspicuous, borne on an inflorescence. The fruit is large, and very rich in sugar.

Cultivation: Prefers a moderately rich soil, full sun or some shade, good drainage. Mulch to improve water retention. Water well in dry season. Suitable for a range of climates. Needs to have an uncrowded position. Slower growth in cooler season, requires less water then.
Sensitive to spider mites.

Propagation: From seed, which may be slow to germinate.

Pink Coral Vine (*Antigonon leptopus*)

Family: Polygonaceae

Alternate Names: Corallita, Chain of Love, Coral Vine, Mexican Creeper, Bride's Tears, Queen's Wreath

Use: To cover walls, fences, trellises, pergolas, and arbors. Used as an urban planting to screen unwanted views. Attracts butterflies, and has edible roots.

Flowers: Large branching racemes (flower stalks), with lacy masses of tiny heart-shaped flowers. In the wild, Pink Coral Vine is usually just that, pink. In cultivation, however, there are white and coral varieties, as well as the purple and pink blend shown below. Long bloom season.

Cultivation: Rapidly grows to 40' in frost-free areas, providing year-round attractive heart-shaped leaves. Needs full sun to part shade in moist well-drained soil and regular feeding during bloom season. Water more in hottest weather. Tolerant of poorer soil and air pollution in urban areas.

Propagation: From seeds. In addition, can root cuttings or transplant volunteer plants from established vines.

Note: Very heat tolerant, not at its best in the coastal climates which are a bit cooler.

Pistachio (*Pistacia*)
Family: Abacarduaceae

Use: Grown as an attractive landscape specimen as well as for its crop of nuts, which are commercially valuable and popular snacks all over the world. Archeological evidence suggests pistachio use as food in Turkey as early as 7000 B.C. The Middle Eastern variety, *P. vera*, is small, growing to about 20', while the Asian variety, *P. chinensis*, grows to 80'.

Flowers: The Asian variety has red flowers and small blue fruits and a long-lasting display of brilliant scarlet, yellow, and purple leaves in cool weather. In other varieties, small brownish green male and female flowers form on separate trees These ripen into reddish fruit borne in clusters much like grapes.

Cultivation: Full sun, deep infrequent watering, let dry to 2" deep. Tolerant of a variety of soils and of high heat, but does not do well with high humidity. Drought resistant. Healthy specimens may live and produce for centuries. Glossy leaves shed dust.

Propagation: By budding or grafting scions onto hardy rootstock.

Plumbago (*Plumbago capensis*)
Family: Plumbaginaceae
Alternate Name: Leadwort

Use: Informal hedge, potted plant, or short-distance climber. Some medicinal uses.
Flowers: Showy clusters of sky-blue 5-lobed flowers resembling phlox. There are deeper blue, orange-red, and white flowers in other varieties.
Cultivation: Well-fertilized garden soil with good drainage. Full sun, protection from cold. Prune to control size and encourage blooms. Drought resistant.
Propagation: By soft-tip cuttings placed in Perlite. Some varieties also spread by suckers.
Note: Moderately poisonous if eaten.

Plumeria (*Plumeria obtusa*)

Family: Apocynaceae
Alternate Names: Graveyard Tree, Frangipani, Temple Flower, Tree of Life

Use: Hawaiian leis, flower arrangements, landscapes. Particularly known for its fragrance.
Flowers: Five petal clusters in white, yellow, shades of pink or red. *P. acuminate* has cream and white flowers; *P. rubra* has red flowers. Flowers appear in clusters from the ends of bare branches, and may continue opening year round. In temperate climates, foliage appears prior to flowers.
Cultivation: Does well in containers, but often seen as shrubs or small trees. Tender to frost and cold. Dwarf varieties grow up to a few feet tall. In tropical climates, full size trees may be as much as 40 feet tall. Pruning recommended to prevent plants from becoming leggy. Full sun needed. Let dry to 2" deep between waterings. Drought tolerant. Goes dormant naturally in cooler climates.
Propagation: Roots easily from cuttings. May continue flowering even when not planted in the ground. Cuttings should be dry at the cut before planting. May also be grown from seed.
Note: The white latex sap is irritating to the skin for some people, and can cause stomach upset if eaten.

Poinsettia (*E. pulcherrima*)
Family: Euphorbiaceae
Alternate Names: Christmas Star, Christmas Flower

Use: Showiest of the *Euphorbia*, millions are sold each year at Christmastime. Also delightful as a garden shrub, growing to 10' with multiple branches. The Aztecs grew them for dye and medicine, as well as for their pleasing color.
Flowers: Branches end in clusters of tiny yellow flowers surrounded by colored bracts, in reds, pinks, and white. Double varieties are available, as are those with variegated bracts.
Cultivation: Part sun, high humidity, rich garden soil with humus, allow soil to dry between waterings. Fertilize every 15 days during bloom period. Prune after flowering; pinch tips to encourage branching. Force bloom by subjecting the plant to 12 hours of darkness each day.
each day. Mildly toxic.

Ponytail Palm (*Beaucarnea recurvata*)
Family: Agavaceae
Alternate Name: Elephant's Foot

Use: Tropical landscape specimen, indoors as a potted plant, also grown as bonsai.
Cultivation: Hardy, forgiving plant for a wide range of conditions. Light well-draining soil, prefers bright sun but tolerates a variety of light conditions. Water lightly, will withstand drought, as the bulb-shaped base stores water. However, it has no defense against over-watering. Slow growing to a height of 30'. Can cut foliage back to base and it will re-grow. Roots will attract mealy bugs if too wet.
Propagation: From seed.
Note: Leaves are knife-sharp.

Portulaca (*Portulaca grandiflora*)
Family: Portulacaceae
Alternate Names: Sun Plant, Rose Moss, Eleven O'Clock, Purslane, Wax Pink

Use: Rock gardens, steep banks or slopes, window gardens, containers. Dense ground cover.
Flowers: Five-petaled waxy flowers about 2" in diameter, some double forms. Bright, vivid colors—pinks, red, yellow, orange, purple, and white. In low light, will not open until mid-day.
Cultivation: Sandy or gravel soil, with leaf mould. Full sun, moderate amounts of water in the hottest season. In cool climates, treat as an annual.
Propagation: From seed, sown in place and lightly covered. Water lightly, and thin when seedlings become crowded, within 2-3 weeks. Will bloom within six weeks. Start seeds under glass where cool.

Pothos (*Raphidophora aurea*)
Family: Araceae
Alternate Names: Devil's Ivy, Marble Queen, *Epipremnum aureum, Scindapsis aureus.*

Use: Attractive climber, with heart-shaped leaves splashed with yellow or white. Indoors, grown as a houseplant, climbing on support to 10'. Most varieties evergreen and variegated.
Flowers: White to near-white. This plant is grown for its showy, variegated leaves.
Cultivation: Indoors, rich organic soil, low to moderate light, regular fertilizer applications, inspect for pests. In gardens and parks, this extremely hardy climber prefers sun to part shade, and a well drained soil, regular water, and protection from wind, which tears the leaves. Gently prune away dead leaves and unattractive arial roots. Contains toxic calcium oxalate crystals, which can be an effective insecticide. This plant is among the most efficient at helping to purify the air.
Propagation: Easily propagated by stem cuttings, leaf bud cuttings, layers, or by division.
Note: Pets eating the leaves will show irritation of the mouth.

Powder Puff (*Calliandra*)
Family: Mimosaceae
Alternate Names: Fairy Duster, Tassel Flower, Lehua Haole

Use: The striking flowers make this an excellent choice for the tropical or subtropical garden. Some of the 200 species of *Calliandra* have utility for firewood, or for high quality animal fodder as well. A natural in wildlife habitat gardens, they provide food for hummingbirds.
Flowers: Most varieties have large pink, red or white tufted pompon-like flowers crowning an open, spreading shrub of about 10' in height. The blooming season is lengthy.
Cultivation: Full sun to part shade. Prefers well-drained soil, but tolerates a variety of planting mediums. Prune for more compact growth. Water heavily during blooming season. New leaves appear brownish, and in some varieties, the leaves fold at sunset. *Calliandra* tolerates sand and seasonal flooding.
Propagation: From ripe seed. Keep warm and moist to germination.

Primavera (*Tabebuia or Cybistax donnell-smithii*)
Family: Bignoniaceae
Alternate Names: White Mahogany, Trumpet-tree, Gold Tree, White Cedar
(The related pink-flowering Amapas,(T. rosea) also a *Tabebuia*, is shown at right, in the wild)

Use: Some varieties of this large rainforest canopy tree grow to 100' in the wild. As a landscape specimen in open spaces, it is primarily grown for its masses of spring-blooming showy yellow flowers. The lustrous wood is of fine cabinet-making quality, creamy white to yellowish rose in color, and beautifully grained.
Flowers: The bright yellow, (or pink, on the *Amapas*), sweetly-scented, trumpet-shaped flowers are spectacular in the late Winter to early Spring. The flowers adorn bare branches, followed by leaves.
Cultivation: Grows rapidly in deep rich soil of warm temperate to tropical climates, both in hot, dry climates, and where there is a rainy season. Deciduous. Grey fruits follow the flowers.
Propagation: By seed, cuttings, or air-layers.

Purple Wreath (*Petraea volubilis*)
Family: Verbenaceae
Alternate Names: Sandpaper Vine, Queen's Wreath

Use: Native to Mexico, this climbing vine is one of the showiest in tropical and subtropical gardens, due to its distinctive and unusual blue and lavender flowers. This woody shrub is at its best against walls, trellised for support.

Flowers: The vine flowers abundantly, with upright or hanging racemes of many two-toned flowers, each blossom about 3/8" long. The corolla is mauve, and is surrounded by longer and narrower lavender blue calyx lobes.

Cultivation: Shelter from strong wind, enrich the soil with organic matter, and provide support. The vine will grow quickly, and its rough surfaced leaves will soon be nearly hidden with beautiful racemes of lavender flowers. Needs well-drained soil, and regular watering. Prune after flowers have passed. Rarely forms fruit in cultivation.

Propagation: By cuttings.

Queen Palm (*Syagrus romanzoffianum*)
Family: Palmae/Arecaceae
Alternate Name: *Cocos plumosa, Arecastrum*

Use: Attractive garden specimen, alone or in casual groupings of three or more. Used to frame views. Frequently used in urban streetscapes. Can be container grown outdoors.
Flowers: An impressive creamy white to yellow flower cluster emerges from large pods during the summer. Flowers are followed by date-sized yellow-orange fruit.
Cultivation: This graceful palm has a smooth, slender, straight grey trunk and grows to 50', topped by a large canopy of feathery drooping plumes. It prefers an enriched sandy soil, but can tolerate a variety of soils and climates as well, and is cold hardy to nearly 25 degrees F. This fast growing palm needs regular water, especially during the dry season.
Propagation: From seed. Or, dig and transplant the volunteer seedlings that appear under the adult plant.
Note: Lives only about 35 years.

Red Leaf Palm (*Chambeyronia macrocarpa*)

Family: Arecaceae
Alternate Names: Jade Flame Thrower, Red Feather Palm, Watermelon Palm, Blushing Palm

Use: Rapidly growing in popularity, this spectacular landscape specimen is best known for its new foliage, which remains an attractive orange-red for as long as 10 days. The contrast with the deep green mature fronds is extremely attractive. Can be grown indoors in cooler climes.
Flowers: Pink to cream flowers, followed by crimson red fruit.
Cultivation: This easy to grow palm reaches a height of 20', with a sturdy trunk and a large green crownshaft with lighter green mottling. The *macrocarpa* (shown) prefers full sun, frequent watering, and well-drained soil. The *hookeri* variety, with its yellow crownshaft, prefers light shade and even more moisture. Both grow well in temperate to tropical climates, and are more cold tolerant than many other palms.
Propagation: From seed. Easy to germinate, grows quickly from seedlings, with slower growth in the more mature plants.

Royal Palm (*Roystonea oleraceae*)
Family: Aracaceae

Use: The 10-12 species of *Roystonea* are stately and striking, often planted as ornamentals. They grow to 80' or more, with a whitish or concrete gray trunk, sometimes swollen in the middle. The dark green leaves can reach 10-12' long. The Caribbee Royal Palm can attain a height of 130', with leaves of 20' in length.

Flowers: Small white fragrant flowers occur in dense drooping clusters of about 2' in length. Flowers of the Cuban Royal Palm have distinctive purple stamens, and purple-colored fruit as well. Flowers of the Caribbee Royal Palm are colorless and their fruit, black.

Cultivation: These fast growing trees require rich soil and plenty of water. They will grow in a wide range of climates, and also make excellent conservatory plants in cooler areas. All species are salt-resistant.

Propagation: All *Roystoneas* are propagated from seed. The small round seeds germinate readily, and sprout within a month.

Royal Poinciana (*Delonix regia*)
Family: Caesalpiniaceae
Alternate Names: Flamboyant Tree, Flame Tree, Peacock Flower

Use: This gorgeous flowering shade tree is sometimes referred to as the showiest flowering tree in the world. It is often planted along streets or in parks. The spreading canopy can exceed the tree's 40' height. Deciduous in climates with dry season.
Flowers: Cluster of flame red flowers 4 -5 " across, consisting of 4 spoon shape scarlet or orange red petals about 3" and one slightly longer petal marked with yellow and white. Leaves are lacy and fern-like, 12-20" long. Seed pods are dark brown, flat, and woody, to 24" long. There is a rare yellow and cream variety as well.
Cultivation: Grows in a variety of well-drained soils with full sun. Tolerates salty conditions. Needs protection from strong winds. May take up to 10 years to flower. The smooth trunk may develop supporting buttresses.
Propagation: By seed, or from semi-ripe tip cuttings.

Russelia (*Russelia equisetiformis*)
Family: Scrophulariaceae
Alternate Names: Firecracker Plant, Fountain Bush, Coral Plant

Use: This native of Mexico is used in masses for garden plantings, and for hanging baskets, rock gardens, and borders. It looks good spilling over from window boxes. Attracts butterflies.
Flowers: Continuously flowers with bright red to coral tubular shaped flowers, borne in loose clusters of around four, each with five lobes at the tip. The "Aurea" variety has yellow-white flowers. Fruit borne rarely in cultivation.
Cultivation: Prefers a rich fertile soil mix and part shade to full sun, regular water and fertilizer. May adapt well to dry conditions. Will grow to a small spreading shrub, about 5' in height and up to 3' in width, mound-like in shape.
Propagation: By air layering where the stems touch the ground, or by cuttings.
Note: This plant is considered invasive in Florida and in parts of Asia.

Sago Palm (*Caryota urens*)
Family: Cycadaceae

Use: Landscape specimens, houseplants. Its starch is a food staple across Southeast Asia and the Pacific. Most common cycad in cultivation in the world.

Flowers: Female produces flowers at 15-20 years, which open, then close and produce seeds if fertilized.

Cultivation: Easy to grow, as it adapts to a wide range of temperature and humidity conditions. Prefers part shade, well-drained, rich soil. Very slow growing, perhaps 3" per year.

Propagation: In the wild, pollination is dependent on wind and birds. Under cultivation, hand pollination produces a high success rate. Sprout in well-drained soil. Prefers to be root-bound when young. Transplant from pots at about three years. Alternatively, can be propagated from "pups", offshoots, or stem cuttings. Allow to harden, or dry, prior to planting. Roots best with bottom heat.

Note: In parts of California, these valuable landscape plants are sometimes chained in place to make them less susceptible to theft.

Salak (*Salacca zalacca*)
Family: Arecaceae
Alternate Name: Snakeskin Fruit

Use: This native of Indonesia is primarily grown for its delicious fruit, something like a cross between apple and pineapple. The creamy colored fruit is covered with a think glossy brown peel that resembles snakeskin. The fruit is eaten raw or cooked with sugar, and stores well.
Flowers: Small inconspicuous flowers produced from the low crownshaft, followed by fruits.
Cultivation: Salak grows best at some elevation, in part shade, and with regular fertilization. It likes ventilation, so light winds are beneficial. The mature long spiny leaves may reach 12-15' on this trunkless palm. The spines, or thorns, on the leaves and on the immature fruit make harvest a task best left to the experienced. In its native Bali, Salak is grown in clay fields, well fertilized and with other plants for shade.
Propagation: Plant seeds in potting mix, keep damp. Will germinate within a month. The seeds can be stored for short periods and remain viable.

Sansevieria (*Sansevieria trifasciata*)
Family: Liliaceae
Alternate Names: Snake Plant, Mother-in-Law's Tongue, Bowstring Hemp

Use: Adds attractive vertical line to flower beds. Plant in clusters in shaded areas, or indoors. Leaves are a fiber source, used in fishing line, hats, mats, shoes, and nets. Can be grown as a container plant indoors, as well, as it tolerates low light levels (although it loses variegation).
Flowers: Has narrow clusters of creamy greenish-white flowers on erect long spires. Very fragrant.
Cultivation: In cool climates, bring potted plants indoors in winter, feed biweekly in the summer to encourage bloom. In warmer climates, plant in shade or semi-shade in sandy compost. Water when the soil is dry; rots easily if over-watered.
Propagation: Divide larger specimens. If leaf cuttings are taken, they will root, but will not display the variegation in color of the mother plant.

Schefflera (*Schefflera actinophylla*)
Family: Araliaceae
Alternate Names: Queensland Umbrella Tree, Octopus Tree, *Brassaia actinophylla*

Use: Most often planted in multiple clumps. Perhaps best known as a popluar indoor potted plant, and as a bonsai specimen. The large erect shrub reaches heights of 30-40'. A smaller variety grows in Hawaii.

Flowers: Flower stems curve and twist, resembling the tentacles of an octopus. Bears upright panicles of small brownish red flowers, followed by round black fruit. The New Zealand variety (*S. digitata*) has green flowers.

Cultivation: Needs part sun to full shade, warm temperatures, high humidity. Indoors, repot every 2 years in a well-drained soil mix. Brown leaf edges indicate overwatering. Inspect for mealy bugs.

Propagation: Can be raised from seed. Alternatively, take large stem cuttings, or air-layer.

Sea Grape (*Coccoloba uvifera*)
Family: Polygonaceae

Use: Tropical garden specimen with dense foliage providing shade and privacy. The timber is used for cabinet making, and the wood is boiled to make a red dye. Other parts of the tree have medicinal application. Fruit is edible, eaten fresh or made into jams and jellies. Often planted along waterways or near beaches, in streetscapes, or for a dense hedge.
Flowers: The spikes of small, greenish white flowers are followed by strings of purple-red fruits, about 3/4" in diameter. The decorative glossy red-veined 8" leaves are shaped like ping pong paddles.
Cultivation: Rapidly grows to 50' high in tropical climates. Highly salt tolerant and grows well in sandy soil. A diffuse, spreading shrub in the wild, Sea Grape is a handsome, vase-shaped tree under cultivation. It is drought tolerant and wind tolerant, and requires full sun to part shade.
Propagation: From seeds or cuttings.
Note: The round, flat leaves were once used as writing paper in schools and missions in early settlements in Mexico.

Senna (*Senna oligophylla*)
Family: Caesalpiniaceae
Alternate Names: *Cassia alata*, Shower Tree

Use: Widespread, attractive tropical and subtropical forest tree, to 50', or large shrub. Occasional garden use as specimen plant. Trees are spreading, providing excellent shade. There are over 200 species. Some medicinal use, and provides tannin. There are temperate climate types.
Flowers: Usually yellow, but also found in red, white, pink, and orange, all have long flowering season and showy displays. Some are fragrant, some rigid spikes, some, long sprays.
Cultivation: Fast growing. There are varieties for nearly every climate. Open space, well-drained soil, full sun. Provides winter color if grown in mass plantings.
Propagation: Grows easily from seeds or cuttings.

Stag-Horn Fern (*Platycerium bifurcatum*)
Family: Polypodiaceae
Alternate Name: Elkhorn

Use: This is a shade-loving epiphyte, and can be grown planted in the fork of a tree, or used indoors, fastened to wood, bark, or cork. The flat, sterile leaves wrap around and secure the plant to the host. Both the flat and the rounded leaves are a pleasing green color.
Flowers: None.
Cultivation: Outdoors, requires part shade. Must have water at the roots, and prefers a humid atmosphere. Plant in equal parts peat and sphagnum moss. One watering technique often used is to let the plant dry, then immerse it in water. Others simply mist often, keeping the humidity high while the planting medium remains only moderately damp. Inspect for mealy bugs, which will appear if overwatered.
Propagation: By spores or offshoots.

Sugar Cane (*Saccharum*)
Family: Gramineae

Use: Grown in the South Pacific for over 6000 years, sugar cane has long been a commercially valuable food crop throughout the tropical world. Cane produces six times the sugar that sugar beets produce, and the sugar cane plant is an extremely efficient photosynthesizer. In addition to sugar production, the plant is used for juice, molasses, candy, dessert, cane fiber, electric power, gasohol and biodiesel fuels, and in the production of alcoholic beverages.

Cultivation: Grows well in tropical and subtropical climates where there is at least 24" of annual rainfall. Planted by hand or machine, the cane can be drip irrigated and fertilized. It forms a dense mat, with the individual canes reaching to 30' in length. Cane can be harvested several times before replanting is required.

Propagation: From cuttings that are about 18" and contain at least one bud. To harvest, the field is set afire, burning the leaves but leaving the water-filled stalk and roots unharmed.

Tabernaemontana divaricata

Family: Apocynaceae

Alternate Names: Fleur d'Amour, *T. coronaria,* Mock Gardenia, Crepe Jasmine, Wax Flower, Pinwheel Flower, Butterfly Gardenia.

Use: While there are more than 300 species of *tabernaemontana* (native of India) growing in the tropics, few are cultivated. *T. divaricata*, however, a beautiful, shapely evergreen shrub rarely more than 6' in height, is widely used in streetscapes and home and commercial gardens. The plant has some medicinal value, *e.g.*, it is a source of ibogaine used in treatment of addiction. The wood is used in perfume and incense.

Flowers: Waxy white delicately fragrant petaled blooms with crepe texture. Long bloom season, followed by half-round fruit. The variety shown is double-flowered; the single variety shows the pinwheel form more clearly.

Cultivation: Full sun to part shade, best in a sheltered spot. Soil should be a mixture of sand, loam, and peat. Semi-tropical or warmer climate. Need regular water, not drought tolerant. Can grow to 10' in ideal conditions. Reliable blooming shrub.

Propagation: From seeds and cuttings.

Note: Roots said to be poisonous.

Teak (*Tectona grandis*)
Family: Verbenaceae
Alternate Names: Teca, Tekka

Use: Tropical hardwood tree grown for its easily worked wood, replete with natural aromatic oils and suitable for boats, outdoor furniture, doors and window frames, and flooring. Termite and fire resistant. Leaves contain a red dye. Some medicinal use for various parts. Grown for its utility for over 2000 years.
Flowers: Numerous small mauve and white flowers develop high in the branches, on terminal cross-branched panicles, appearing after the rains. Flowers are followed by dry brown fruits.
Cultivation: Grows best in dry, hilly terrain to 3000' altitude in tropical to subtropical climates. Needs good drainage, and will grow relatively quickly in ideal conditions, reaching a maximum height of 150-160', with a trunk 6-8' in diameter. Under plantation conditions, can produce a clear 100' stem with high timber yield.
Propagation: By seed naturally, by shoots in cultivation. Experimentally, from one year old stem cuttings, tissue culture, and clonal propagation. Due to the disappearance of natural old-growth teak forests, much effort is underway to increase the efficiency of plantation cultivation for this valuable wood source.

Teddy Bear Palm (*Dypsis lastelliana, d. leptocheilos*)
Family: Aracaceae
Alternate Names: Col rouge, Redneck Palm, Majestic Palm

Use: This attractive, rare, palm is seen as a specimen in tropical landscape design, and can be used, due to its height, as an unusual frame for an entryway. The crownshaft is striking in that it is covered with fuzzy rust-brown hairs, soft to the touch, giving it the "teddy bear" name (see photo at top right). The remaining part of the trunk is a very nice silver green, with white rings.
Flowers: Inconspicuous, borne on an inflorescence.
Cultivation: Adaptable to many soil types if well drained. Requires bright sun. Irrigate in dry season, protect from wind. Will adapt to shade.
Propagation: From seed, which germinates in 1-2 months.

Thevetia (*T. aurantiaca*)
Family: Apocynaceae
Alternate Names: Yellow Oleander, Be-Still Tree, Lucky Nut

Use: With their gorgeous golden or orange trumpets, these are beautiful as landscape specimens or as hedges. They are, however, poisonous in all parts. They can be left in their bushy natural shape or pruned to a neater tree shape. The eight species are originally from the Americas, but are cultivated worldwide in warm climates.
Flowers: The variety seen here is orange-flowered, but others bear yellow clusters (*T.thevetioides*) or those of golden hue (*T.peruviana*).The flowers are somewhat fragrant and are followed by poisonous fruits.
Cultivation: Well-watered, ordinary garden soil, moderate to hot climates. Related to oleander, these thrive where oleander does well. Tolerates mild frost. Support stakes are helpful. *T. peruviana* is called the Be- Still Tree for the constant movement of its delicate leaves. These plants can withstand some frost if sheltered.
Propagation: From seed or cuttings.

Thunbergia (*Thunbergia grandiflora*)

Family: Acanthaceae
Alternate Names: Blue Sky Flower, Glory Vine, Blue Trumpet Vine, Clockvine

Use: Vigorous climber, also sold in hanging baskets. Available in perennial and shrub form as well. All told, there are 100-200 *Thunbergia* species, many grown for their large flowers. Often planted to grow upward on trellises.
Flowers: This species has delicate blue flowers; other species have flowers of white, purple, or gold (black-eyed Susan). *T. grandiflora* has curtains of flaring tubular flowers, 3" in diameter.
Fruit is not usually formed in cultivation.
Cultivation: For best bloom, needs full sun to part shade, consistently moist, well-drained soil that contains organic matter.
Propagation: Vines may be started from seeds or cuttings. If planted from seed, there may be color variations.

Tradescantia (*T. flaminensis*)
Family: Commelinaceae
Alternate Name: Zebrina, Inchplant, Purple Wandering Jew, *Zebrina pendula*

Use: Forms dense garden mat, so most often used for ground cover. Also valuable as a decorative hanging plant. The upper side of the leaves is silver green with a purple stripe and purple edges, while the underside is bright pinkish-purple. This is a succulent from the herb family. In shade, the leaves are not as colorful.

Flowers: Small, pink or magenta, three-petaled, enclosed by two leaflike bracts.

Cultivation: This Mexican native does well in full sun to part shade. It requires little water, and will grow in most any garden soil. This plant is so adaptable that it is considered invasive in many areas. If it gets too much sun, there may be bleaching of the leaf color. Fruit is not often formed in cultivation.

Propagation: From seeds, or let it root from nodes.

Traveler's Palm (*Strelitzia ravenala*)
Family: Musaceae
Alternate Name: Traveler's Tree

Use: Primarily as a landscape specimen, and its large fan is also used for shade. The name derives from the bases of the leaf sheaths, which contain water and have been of great emergency value to thirsty travelers. It is not actually a palm, but rather a member of the banana family. The sap is rich in sugar, and nutritious. Reaches 30'-60'.
Flowers: The small creamy white flowers are borne from multiple boat-shaped bracts among the stems, that greatly resemble the Bird of Paradise flower, which is also a member of the banana family.
Cultivation: Tolerates sand or clay with good drainage, thrives in rich soil. Full sun or part shade, monthly feeding. Can tolerate dry spell up to 3 months, but does best with regular watering.
Propagation: From seed or by dividing the new stems which it sends up from the soil.

Tree Philodendren (*P. selloum*)
Family: Araceae
Alternate Names: Lace Tree Philodendron, Split Leaf Philodendron

Use: As a large container plant in public spaces, or as a specimen shrub on an expanse of lawn. Can be used under trees, or against a pool wall, or massed in semi-shady areas. They are helpful in cleaning toxins from the air.
Flowers: Insignificant white flowers. Grown for the attractive leaves.
Cultivation: One of some 200 species, this plant is grown indoors as a potted plant in cooler climates. In tropical areas, plant in part sun, with more light required by the variegated cultivars. Grows best in moist well drained fertile soil. Trim away discolored lower leaves, water during dry spells. May reach 15' in height and width. Will grow in very shady places.
Propagation: From cuttings. Place shoots in potting medium, roots in 4-8 weeks. Or, start from seed placed in a moist peat moss/ sand mixture.
Note: Tree Philodendron has an oxalic sap, which is an effective natural pesticide, especially effective against spider mites.

Triangle Palm (*Dypsis decari*)
Family: Arecaceae
Alternate Name: *Neodypsis decari*

Use: Bold and formal, this fast growing palm makes an excellent accent plant in open space. Its three planed leaf base provides the name, for the leaves emerge from a triangular formation. The leaves are a striking blue grey, growing straight upward, arching downward at the tips.
Flowers: Yellow to green borne on a branched inflorescense. Produces round 1" black fruit.
Cultivation: Full sun, regular water. Warm to tropical climate, drought tolerant. Grows 20' typically, but native trees have reached heights of 50'.
Propagation: From seed, which germinates in about 1 month.
Note: The palm in the photograph in the center is a true type; the palm at the lower left is crossbred with *Areca*.

Tulipan (*Spathododea campanulata*)
Family: Bignoniaceae
Alternate names: African Tulip Tree, *Spathodea nilotica,* Sorcerer's Wand, Flame of the Forest, Tulip Tree, Fountain Tree

Use: Planted as a specimen tree, for shade, or along city streets or in parks. Used throughout the warmer climates of the world. Grows to 50' or more in the wild.
Flowers: The beautiful, bright orange-scarlet flowers are 4" in diameter and are lined with yellow. They appear in large racemes at branch ends, and open a few at a time, with the process lasting several months, providing a lengthy display. "Kona Gold" has yellow flowers.
Cultivation: Tolerates many soil types, but does best in rich soil. Seen in moist habitats below 3000'. Will grow to as much as 80' in sheltered ravines. Full sun and regular watering for best flowering, but will tolerate drought. Prune heavily after any freeze.
Propagation: From seed, which are contained in woody capsules. They are winged, and the wind carries them afar, naturalizing the trees. Also can be propagated by tip cuttings and by root cuttings, or suckers.
Note: The trees may become hollow with age posing a hazard of breaking and falling branches.

Vinca (*Vinca major*)

Family: Apocinaceae

Alternate Names: Madagascar Periwinkle, Vinca rosea, Catharanthus, Lochnera rosea

Use: Borders, window boxes, indoor potted plant, beds. Some medicinal use, particularly in the treatment of diabetes and cancer. Goes dormant in winter in Northern climes.

Flowers: Almost everblooming. Large, flat 5-petaled flowers in rose pink or white on long leafy stems. Shrub-like perennial or dwarf annual.

Cultivation: Good light potting compost. Full sun or half shade, light regular watering.

Propagation: From seed or by soft cuttings. Set seedlings out in well-drained, enriched garden soil. In cool climates, sow seed indoors in mid-winter.

Water Lilies (*Nymphaea*)
Family: Nymphaeaceae

Use: These aquatic perennials are hardy and easy to grow. They add color and grace to any landscape water feature, with their handsome floating foliage and beautiful flowers. Many hybrids have been developed from the wild species; there are now over 1600 named species. A selection of those seen in resorts in Mexico are photographed, below.

Flowers: Water Lily flowers come in nearly all colors, and a dazzling array of sizes, from the tiny, quarter-sized varieties to those with leaves three feet across that will support a standing person.

Cultivation: The plants should be set into pots filled with any heavy garden soil so that the planting medium will not float away. Fertilize regularly, and provide at least 5 hours of sunshine daily. The water level should be no more than 6" above the top of the pot. Keep the water fairly clean, and groom the plants by removing dead leaves and pests.

Propagation: Tropical water lilies are grown from tubers produced at the end of the growing season. Hardy water lilies are increased through division of rhizomes. Both can also be grown from seed; dry the seed for Tropical types; keep the seed for Hardy types in water. Plant Tropicals level in the container, Hardy types at a 45 degree angle against the side of the pot.

Wheat Celosia (*C. spictata*)
Family: Amaranthaceae
Alternate Names: Feathered Amaranth, Flaming Spears

Use: One of the 60 or so species of Celosia, this variety, in masses, is excellent for beds, borders, and edging. Also makes a good dried flower if care is taken to avoid shattering the flowerhead. Some seed providers report that celosia has historically been used for medicinal purposes as well.

Flowers: There are hundreds of tiny flowers packed in dense flowerheads which stand above the foliage. These feather-like spikes can be found in rose, purple, yellow, red, orange, and crimson.

Cultivation: These heavily branched annual plants can reach 24" and thrive in hot, dry conditions in full sun. They grow well in average well drained garden soil, and bloom best with regular fertilizer.

Propagation: From very fine seed. Cover lightly and keep warm. Water well but let dry between waterings. Plant 12" apart. Germinates in 10 days, blooms in 70-100 days.

Note: The name Celosia comes from a Greek word that means " burning".

White Mangrove (*Luguncularia racemosa*)
Family: Rhizophoraceae

Use: There are about 60 species of this evergreen tree and shrub. All form dense thickets in tidal creek estuaries, with the roots catching soil and debris carried by tide. This process extends the shoreline over many years. The bark yields tannin, the wood, charcoal, and timber. Fishermen use the fruit of the Freshwater Mangrove, *Barringtonia acutangula,* for net floats, and grate the seed to stun and catch fish. Mangroves provide a natural habitat for birds and fish, and are critical to the local ecosystem.

Flowers: Red Mangrove flowers are pale yellow clusters, those of *B. acutangula,* bright pink and white puffballs that bloom at night.

Cultivation: Often seen as small trees, to 20', but can grow to 80' in the tropics. Although found at the shoreline and along tributaries, they also can be grown as a garden specimen. Requires at least a subtropical climate.

Propagation: The seeds germinate while attached to the tree, then drop and float. When they fall into mud, they immediately begin growth.

Yellow Shrimp Plant (*Pachystachys lutea*)
Family: Acanthaceae

Alternate Names: Flaming Golden Candles, Lollipop Plant, Camarones (shrimps)

Use: There are 12 tropical American shrub species of Pachystachys. It makes a marvelous bedding plant with East or West exposure. In cooler climates, best as a container plant, especially for window boxes and summer urns. This bushy flowering plant, typically 5' tall, can grow twice that size under ideal conditions, and tends to be leggy. Smaller foliage or flowering plants in front increase its attractiveness. It is neither tall enough nor strong enough to function as a hedge, but can be an eye-catching border in front of hedges.

Flowers: Bright yellow spikes of bracts provide months of color. The 2-lipped long white flowers will bloom for months as well.

Cultivation: Filtered sun, part sun, or shade. High humidity, and moist soil, but once established, this plant's water requirements are low. Grows in any standard garden soil or potting mix. Fertilize bi- weekly, hand prune for neat appearance. In the right conditions, it will produce numerous stems up to 3' in height, each with the golden spike at the top.

Propagation: From tip cuttings.

Yucca (*Yucca gloriosa*)
Family: Liliacea

Use: Adaptable desert-dweller, makes an attractive feature plant in many types of gardens. May be used as an accent plant, isolated on lawns, or placed in deep containers. The sharp leaves are best kept away from paths, however.

Flowers: The creamy white panicles of bell-shaped flowers may appear at any time during warm weather. The flower display is spectacular, with some clusters of the densely packed flowers reaching more than 3' in length. Flowers are edible.

Cultivation: Needs average water supply, partial sun, adequate room, at least an area 10' X 10'. Will branch out. Widespread in Mexico, it has also been cultivated in southeastern US. Use care; leaves and tips are sharp.

Propagation: By division of offsets. Detach offsets from old stems, root in sand/loam mixture. Alternatively, sever whole heads of foliage from mature plants below the leaf mass, and root in a soil-gravel mix.

Zamia maritima (*farfurecea*)
Family: Cycad
Alternate Name: Cardboard Palm

Use: Landscape specimen, attractive plant for large containers in public spaces, balconies, patios, also grown as bonsai.
Flowers: Inconspicuous/none.
Cultivation: Tolerates dry season, various soil conditions, hardy in full sun to part shade. Salt tolerant, grows at sea level and low altitudes. Can be established in rocky or sandy soil. Grows to 4-5', with many branches from an underground or emergent trunk. May reach a spread of 6' in good growing conditions. Widely used for its hardiness and attractiveness.
Propagation: Both males and females produce cones. Sow seeds from cones, cleaned and dried, promptly.
Note: All parts poisonous. Seeds may be harmful to pets. This "living fossil" plant has survived since the time of the dinosaurs. The leaves are slightly fuzzy, and have a cardboard feel. In shade, leaves will be further apart than in full sun.

Index
(Items in bold are in alphabetical order in the text)

Acalypha
Acalypha hispida, see *Acalypha*
Adenium obesum, see Desert Rose
African Tulip Tree, see Tulipan
Agave Americana, see Century Plant
Agave tequila weber azul, see Blue Agave
Airplant, see Kalenchoe
Alexander Palm, see King Palm
Allamanda
Allamanda cathartica, see *Allamanda*
Amapas, see Primavera
Amarantha, see *Acalypha*
American Oil Palm, see Mexican Oil Palm
Angels Trumpet, see Datura
Anthurium
Antigonon leptopus, see Pink Coral Vine
Araucania exelsa heterophyla, see Norfolk Island Pine
Archontophoenix alexandrae, see King Palm
Arecastrum, see Queen Palm
Artocarpus altilis, see Breadfruit
Arundinaria, see Bamboo
Attalea cohune, see Mexican Oil Palm
Bamboo
Banana
Banjo Fig, see Fig
Banyan, see *Ficus Benjamina*
Barbados Pride, see Dwarf Poinciana
Barbary Fig, see Nopal
Bauhinia variegata, see Orchid Tree
Be-Still Tree, see Thevetia
Beaucarnea recurvata, See Ponytail Palm
Beefsteak Plant, see *Acalypha*
Benjamin's Fig, see Ficus Benjamina
Bird of Paradise
Bishop's Egg, see *Fagraea berteroana*
Bismark Palm
Bismarckia nobilis, see Bismark Palm

Blue Agave
Blue Sky Flower, see Thunbergia
Blue Trumpet Vine, see Thunbergia
Boat Lily, see Oyster Plant
Bottle Palm
Bowstring Hemp, see Sansevieria
Bougainvilla
Brasseria actinophylla, see Schefflera
Breadfruit
Bride's Tears, see Pink Coral Vine
Blushing Palm, see Red Leaf Palm
Bromeliads
Brugmansia suaveolens, see Datura
Buddha's Lamp, see Mussaenda
Bursera simaruba, see Gumbo Limbo
Bush *Allamanda*, see *Allamanda*
Butterfly Flower, see Orchid Tree
Butterfly Gardenia, see Tabernaemontana
Butterfly Palm
Button Mangrove, see Buttonwood
Buttonwood
Caesalpinia pulcherrima, See Dwarf Poinciana
Calabash
Caladium
Caladium hortelanum, see Caladium
Calliandra, see Powder Puff
Camarones, see Yellow Shrimp Plant
Candleabra Cactus
Canna Lily
Cardboard Palm, see Zamia Maritima
Carica papaya, see Papaya
Caryota, see Fishtail Palm
Caryota rumphiana, see Fishtail Palm
Caryota urens, see Sago Palm
Cassia alata, see Senna
Catharenthus, see Vinca
Celosia
Celosia argentia "cristata", see Celosia
Celosia spictata, see Wheat Celosia
Century Plant
Chambeyronia macrocarpa, see Red Leaf Palm
Chain of Love, see Pink Coral Vine
Chilean Jasmine, see Mandevilla
Chinese Wool Flower, see Celosia

Christmas Cheer, see Jellybean Plant
Christmas Flower, see Poinsettia
Christmas Palm, see Manilla Palm
Christmas Star, see Poinsettia
Clockvine, see Thunbergia
Coccoloba uvifera, see Sea Grape
Coconut Palm
Coccos nucifera, see Coconut Palm
Cocos plumosa, see Queen Palm
Codiaeum variegatum, see Croton
Coffee
Coffea Arabica, see Coffee
Cohune Palm, see Mexican Oil Palm
Col rouge, see Teddy Bear Palm
Conocarpus erectus, see Buttonwood
Copper Leaf, see *Acalpha*
Coral Plant, see Jatropha
Coral Vine, see Pink Coral Vine
Corallita, see Pink Coral Vine
Cordyline
Cordyline terminalis, see Cordyline
Corozo Palm, see Mexican Oil Palm
Crane Flower, see Bird of Paradise
Crane's Bill, see Bird of Paradise
Crepe Jasmine, see Tabernaemontana
Crescentia cujete, see Calabash
Crinum augustum
Crossandra
Crossandra infundibuliformis, see Crossandra
Croton
Cybistax donell-smithii, see Primavera
Cyperus papyrus, see Papyrus
Datura
Dawn Flower, see Morning Glory Tree
Delonix regia, see Royal Poinciana
Desert Azalea, see Desert Rose
Desert Rose
Devil's Ivy, see Pothos
Dipladenia, see Mandevilla
Dracaena
Dracaena concinna, see Dracaena
Dracaena marginata "tricolor", see Dracaena
Dwarf Poinciana
Dypsis decari, see Triangle Palm
Dypsis lastelliana, see Teddy Bear Palm
Dypsis leptocheilos, see Teddy Bear Palm
Dypsis lutescens, see Butterfly Palm

Egyptian Paper Plant,
  see Papyrus
Elephant's Foot, see
  Ponytail Palm
Eleven O'clock, see Portulaca
Elkhorn, see Staghorn Fern
Epipremnum aureum, see Pothos
Euphorbia ingens triloba, see
  Candelabra Cactus
Euphorbia pulcherrina,
  see Poinsettia
Fagraea berteroana
Fairy Duster, see Powder Puff
Farfurecea, see Zamia Maritima
Feathered Amaranth,
  see Wheat Celosia
Ferns
Ficus, see *Ficus Benjamina*
Ficus Benjamina
Ficus carica, see Fig
Ficus elastica
Ficus lyrata, see Fig
Fiddle-leaf Fig, see Fig
Fig
Fire Dragon Plant, see
  Acalypha
Firecracker Flower, see
  Crossandra
Fishtail Palm
Flag Bush, see Mussaenda
Flamboyant Tree, see
  Royal Poinciana
Flame of the Forest,
  see Tulipan
Flame of the Woods, see
  Ixora
Flame Tree, see Royal
  Poinciana
Flame Vine
Flaming Spears, see
  Wheat Celosia
Flaming Golden Candles,
  see Yellow Shrimp Plant
Flamingo Flower, see *Anthurium*
Fountain Tree, see Tulipan
Foxtail Palm
Geraniums of the South,
  see Caladium
Frangipani, see Plumeria
Ginger
Glory Vine, see Thunbergia
Gold Tree, see Primavera
Golden Cane Palm see
  Butterfly Palm
Golden Cockade, see Guzmania
Golden Feather Palm,
  see Butterfly Palm
Golden Trumpet, see *Allamanda*
Good Luck Palm, see
  Licuala spinosa
Graveyard Tree, see Plumeria
Guabaya, see Guava
Guava
Gumbo Limbo
Guzmania
Guzmania lingulata minor,
  see Guzmania
Halfmens
Handkerchief Plant, see
  Mussaenda

Heliconia
Heliconia caribeae, see
  Heliconia
Hibiscus
Hibiscus rosa-sinesis, see
  Hibiscus
Huanacaxtle
Hyophorbe lagenicaulis,
  see Bottle Palm
Huernia
Huernia hystrix, see Huernia
Huernia schneideriana,
  see Huernia
Impala Lily, see Desert Rose
Inchplant, see Tradescantia
Indian Mulberry, see
  Morinda Noni
Indian Rubber Tree, see
  Ficus elastica
Indian Shot, see Canna Lily
Ipomoea arborescens, see
  Morning Glory Tree
Iverea pinnata, see
  Kalenchoe
Ixora
Ixora chenensis, see Ixora
Jade Flame Thrower, see
  Red Leaf Palm
Jatropha
Jellybean Plant
Jungle Flame, see Ixora
Jungle Geranium, see Ixora
Kalenchoe
Kalenchoe brasiliensys,
  see Kalenchoe
Kalenchoe pinnata, see
  Kalenchoe
King of Fruits, see Mango
King Palm
Lace Tree Philodendron,
  See Tree Philodendron
Lantana
Lantana camera, see Lantana
Leadwort, see Plumbago
Lehua Haole, see Powder Puff
Licuala grandis
Licuala spinosa
Little Boy Flower, see
  Anthurium
Liver Leaf, see Oyster Plant
Llamarada, see Flame Vine
Lobster Claw, see Heliconia
Lochnera rosea, see Vinca
Lolipop Plant, see
  Yellow Shrimp Plant
Lucky Nut, see Thevetia
Luguncularia racemosa,
  See White Mangrove
MacArthur Palm
Madagascar Palm, see
  Butterfly Palm
Madagascar Periwinkle,
  see Vinca
Majestic Palm, see
  Mexican Oil Palm
Majestic Palm, see
  Teddy Bear Palm
Mandevilla
Mangrove Fan Palm, see
  Licuala spinosa

Mangifera indica, see Mango
Manilla Palm
Mango
Marble Queen, see Pothos
Maritime Pine
Mexican Creeper, see
  Pink Coral Vine
Mexican Fan Palm
Mexican Geranium, see Ixora
Mexican Love Vine, see
  Mandevilla
Mexican Loveplant, see
  Kalenchoe
Mexican Oil Palm
Milk and Wine Lily, see
  Crinum augustum
Miracle Leaf, see Kalenchoe
Mock Gardenia, see
  Tabernaemontana
Money tree, see Dracaena
Moon Flower, see
  Datura, also Morning
  Glory Tree
Morinda citrofolia, see
  Morinda Noni
Morinda Noni
Moses in the Bulrushes,
  see Oyster Plant
Mother-in-Law's Tongue,
  see Sansevieria
Musa, see Banana
Morning Glory Tree
Mussaenda
Mussaenda frondosa, see
  Mussaenda
Neodypsis decari, see
  Triangle Palm
Nerium oleander, see Oleander
Nopal
Norfolk Island Pine
Obake, see *Anthurium*
Octopus Tree, see Schefflera
Nymphaea, see Water Lilies
Oleander
Orange-flowered Stephanotis,
  see Flame Vine
Opuntia ficus-indica, see
  Nopal
Orchid Tree
Orchids
Organ Pipe Cactus
Oyster Plant
Pachypodium, see Halfmens
Pachystachys lutea, see
  Yellow Shrimp Plant
Painkiller Tree, see
  Morinda Noni
Palatte Flower, see *Anthurium*
Palo Blanco, see
  Morning Glory Tree
Papaya
Paper Flower, see *Bougainvilla*
Papillilo, see Gumbo Limbo
Papyrus
Parota, see Huanacaxtle
Parrot Flower, see Heliconia
Peach of the Tropics,
  see Mango
Peacock Flower, see
  Dwarf Poinciana

Peacock Flower, see
    Royal Poinciana
Petraea volubilis, see
    Purple Wreath
Philodendron selloum, see
    Lace Tree Philodendron
Philodendrons
Phoenix humilis royale,
    see Pygmy Date Palm
Phoenix roebelinii, see
    Pygmy Date Palm
Pink Coral vine
Pink Jellybeans, see
    Jellybean Plant
Pino maritime, see Maritime
    Pine
Pinwheel Flower, see
    Tabernaemontana
Pistachio
Pistacia, see Pistachio
Platycerium bifurcatum,
    see Staghorn Fern
Pleomele marginata,
    see Dracaena
Plumbago
Plumbago capensis, see
    Plumbago
Plumeria
Plumeria obtuse, see   Plumeria
Pothos
Powder Puff
Poinsettia
Ponytail Palm
Pork and Beans, see
    Jellybean Plant
Portulaca
Portulaca grandiflora,
    see Portulaca
Psidium guajava, see Guava
Prickly Pear Cactus, see Nopal
Primavera
Pteridophytes, see Ferns
Ptychosperma macarthurii,
    see Macarthur Palm
Pua-kenikeni, see
    Fagraea berteroana
Purple Wandering Jew,
    See Tradescantia
Purple Wreath
    Purslane, see Portulaca
Pygmy Date Palm
Pyrostegia venusta, see
    Flame Vine
Queen Palm
Queen's Wreath, see
    Purple Wreath
Queensland Umbrella
    Tree, see Schefflera
Rain Tree, see Mexican
    Oil Palm
Rainbow Tree, see Dracaena
Raphidophora aurea, see Pothos
Red Cockade, see Guzmania
Red Feather Palm, see
    Red Leaf Palm
Red Hot Cat's Tail, see
    Acalypha
Red Leaf Palm
Red Sage, see Lantana
Redneck Palm, see

Teddy Bear Palm
Rhoeo discolor, see Oyster Plant
Rose Mallow, see Hibiscus
Rose Moss, see Portulaca
Rose of China, see Hibiscus
Rose of Sharon, see Hibiscus
Rosebay, see Oleander
Rosella, see Hibiscus
Royal Palm
Royal Poinciana
Roystonea oleraceae, see
    Royal Palm
Rubber Plant, see *Ficus elastica*
Sabie Star, see Desert Rose
Ruffled Fan Palm, see
    Licuala grandis
Ruffled Lantan Palm, see
    Licuala grandis
Saccharum, see Sugar Cane
Sago Palm
St. Thomas Tree, see
    Orchid Tree
Salacca zalacca, see Salak
Salak
Sandpaper Vine, see Purple
    Wreath
Sansevieria
Sansevieria trifasciata, see
    Sansevieria
Schefflera
Schefflera actinophylla,
    see Schefflera
Scindapsis aureus, see Pothos
Sea Grape
Sedum Rubotinctum "Aurora",
    see Jellybean Plant
Senna
Senna oligophylla, see
    Senna
Shower Tree, see Senna
Shrub Althaceae, see Hibiscus
Shrub Verbena, see Lantana
Snake Plant, see Sansevieria
Snakeskin Fruit, see Salak
Sorcerer's Wand, see Tulipan
Spathe Flower, see *Anthurium*
Spathododea campanulata, see
    Tulipan
Spathodea nilotica, see Tulipan
Spiny Linuala, see Licuala
    spinosa
Split Leaf Philodendron,
    see Tree Philodendron
Staghorn Fern
Star Pine, see Norfolk Island
    Pine
Stenocereus thurberi, see
    Organ Pipe Cactus
Strelitzia ravenala, see
    Traveler's Palm
Strelitzia regina, see
    Bird of Paradise
Sugar Cane
Sun Plant, see Portulaca
Swamp Lily, see *Crinum
    Augustum*
Syragrus romanzoffianum,
    See Queen Palm
Tabebuia, see Primavera
Tabebuia rosea, see *Primavera*

Tabernaemontana
Tabernaemontana coronaria,
    see Tabernaemontana
Tabernaemontana divaricata,
    see *Tabernaemontana*
Tassel Flower, see Powder Puff
Teak
Teca, see Teak
Tectona grandis, see Teak
Teddy Bear Palm
Tekka, see Teak
Temple Flower, see Plumeria
Thevetia
Thevetia aurantiaca, see
    Thevetia
Thunbergia
Thunbergia grandiflora,
    see Thunbergia
Tourist Tree, see Gumbo Limbo
Tradescantia
Trandescantia fluminensis,
    See Tradescantia
Traveler's Palm
Traveler's Tree, see Traveler's
    Palm
Tree Gourd, see Calabash
Tree of Life, see Plumeria
Tree Philodendron
Triangle Palm
Trumpet-tree, see Primavera
Tulip Tree, see Tulipan
Tulipan
Vanuata Fan Palm, see
    Licuala grandis
Veitchia merrillii, see
    Manilla Palm
Vinca
Vinca major, see Vinca
Vinca rosea, see Vinca
Washingtonia robusta, see
    Mexican Fan Palm
Water Lilies
Watermelon Palm, see
    Red Leaf Palm
Wax Flower, see
    Tabernaemontana
Wax Pink, see Portulaca
Weeping Fig, see *Ficus
    Benjamina*
West Indian Birch, see
    Gumbo Limbo
Wheat Celosia
Wild Sage, see Lantana
Wine Palm, see Fishtail Palm
White Cedar, see Primavera
White Mahogany, see
    Primavera
White Mangrove
Wodyetia bifurcata, see
    Foxtail Palm
Yellow Oleander, see Thevetia
Yellow Palm, see Butterfly Palm
Yellow Sage, see Lantana
Yellow Shrimp Plant
Yucca
Yucca gloriosa, see Yucca
Zamia Maritima
Zebrina, see Tradescantia
Zebrina pendula, see Tradescantia
Zinziber, see Ginger

Bibliography*

Blomberg, Alec & Tony Rodd. <u>Palms of the World.</u> North Ryde, NSW, Australia, Angus & Robertson Pub., 1982.

Brockman, C. Frank. <u>Trees of North America.</u> N. Y., Golden Press, 1986.

Chan, Elisabeth. <u>Handy Pocket Guide to Tropical Plants.</u> Singapore, Periplus Editions, 2003.

Cronin, Leonard. <u>Key Guide to Australian Palms, Ferns and Allies.</u> Frechs Forest, NSW, Australia. Reed Books Pty Ltd.1989.

Macoboy, Stirling. <u>What Flower is That?</u> Edison, New Jersey, Chartwell Books, Inc., 2000.

Macoboy, Stirling. <u>What Tree is That?</u> Sydney, Australia, Weldon Publishing, 1979., Rev. Ed., N. Y., Random House (Crescent Books), 1991.

Moran Brandies, Monica, and the Editors of Sunset Books. <u>Landscaping with Tropical Plants.</u> Menlo Park, California, 2004.

Nellis, David W. <u>Seashore Plants of South Florida and the Caribbean.</u> Sarasota, Fla., Pineapple Press, Inc., 1994.

Perrero, Laurie. <u>The World of Tropical Flowers.</u> Miami, Florida, Windward Publishing, Inc., 1976.

Riffle, Robert Lee & Paul Craft. <u>An Encyclopedia of Cultivated Palms.</u> Portland, Timber Press, 2003.

<u>Simon and Schuster's Complete Guide to Plants and Flowers.</u> Frances Perry, ed. N. Y., Simon & Schuster, 1974.

Warren, William. <u>The Tropical Garden.</u> London, Thames and Hudson Ltd., 1991.

Welland, Frances. <u>Place That Plant.</u> Bath, UK, Parragon Publishing, 2001.

Whistler, W. Arthur. <u>Tropical Ornamentals, a guide.</u> Portland, Oregon, Timber Press, 2000.

* In addition, a large number of internet sites were visited during the research for this book. Many are helpful, some are erroneous; the reader is advised to use caution .